Back to the Well

Also by Marq de Villiers

Our Way Out
Dangerous World
Timbuktu
Witch in the Wind
Windswept
A Dune Adrift
Sahara
Water
Into Africa
Blood Traitors
Down the Volga
The Heartbreak Grape
White Tribe Dreaming

MARQ de VILLIERS

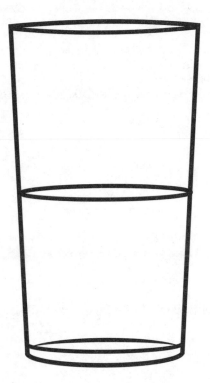

BACK TO THE WELL

Rethinking the Future of Water

Edited by Susan Renouf.
Cover and page design by Chris Tompkins.
Lyrics from "Dress Rehearsal Rag" by Leonard Cohen
copyright © 1970 Stranger Music, Inc. Used by permission.
Printed in Canada.
10 9 8 7 6 5 4 3 2 1

Library and Archives Canada Cataloguing in Publication

De Villiers, Marq, 1940-, author
Back to the well : rethinking the future of water / Marq de Villiers.

Includes bibliographical references and index.
Issued in print and electronic formats.
ISBN 978-0-86492-075-1 (bound).--ISBN 978-0-86492-802-3 (epub).--
ISBN 978-0-86492-572-5 (mobi)

1. Water-supply. 2. Water-supply--Forecasting. I. Title.

HD1691.D46 2015   333.91   C2015-901855-2
C2015-901856-0

We acknowledge the generous support of the Government of Canada, the Canada Council for the Arts, and the Government of New Brunswick.

Nous reconnaissons l'appui généreux du gouvernement du Canada, du Conseil des arts du Canada, et du gouvernement du Nouveau-Brunswick.

Goose Lane Editions
500 Beaverbrook Court, Suite 330
Fredericton, New Brunswick
CANADA E3B 5X4
www.gooselane.com

By means of water we give life to everything.
— The Quran

Oh! I'm so glad!
— Pollyanna

Contents

Introduction
A Few Assertions

An Assertion

Have we really reached peak water, the point at which the renewable (and safe) supply is forever outstripped by unquenchable demand? Or was Pollyanna right to be glad, and our water woes can be fixed by clever management?

Sure, it's true, as a recent UN report put it, that global water use has grown at more than twice the rate of the world's population for the last century, and it's true that we are drawing down, or overdrafting, many of our water resources at a rate that is unsustainable, and it's true that we are still polluting water that we should have cleaned up decades ago. But no, we're not necessarily doomed, and this book will explain why.

A Second Assertion

Looked at one way, the "water crisis" is global. Looked at more closely, the crisis splits into two intersecting and overlapping crises: the crisis of supply (shortages) and the crisis of contamination (pollution). Looked at more correctly, there are no global-scale crises. Instead, there are a thousand smaller regional and river-basin crises, only some of which intersect or overlap with the others, and there are, in fact, still many places without crises at all. This makes water problems more tractable — easier to solve, not harder. This book will delve into that too.

And a Third

If we can avoid the most deeply irrelevant ideological quarrels to which the water world is so prone (the notion of a callous Big Water cartel that would reserve clean water for the rich, paranoia about bulk water transfers from "us" to "them," polemics against dams, quarrels over whether water should be defined as a human right), a wide range of techniques will take us very close to solution. Except in spots, of course. Some places are harder. We will need to examine those.

And One More

What's wrong with privatization? No, really... what's wrong with it?

And a Victim Impact Statement

A Spanish hydrologist, Ramón Llamas Madurga, a contrarian by nature, not long ago annoyed a UNESCO water conference (called Water: The Looming Crisis) by saying, "Why are we always proclaiming that crisis is looming, and yet it never actually looms? With water, we have seemingly been standing on the edge of the cliff of catastrophe for decades, yet we never seem to fall into the abyss."

He was suggesting, to grumbles from the audience, that the notion of "crisis" might be alarmist. Alarmist and, therefore, unhelpful.

But he was looking, I think, in the wrong places, and at the wrong scale — he was looking for a single, identifiable, overweening global crisis. In many parts of the world, there is no such thing. In others, they have *already* stumbled over the edge.

For some, it may be too late. But for others, there is still time for a lifeline.

People who write about population policy sometimes forget that "population" is just people in the aggregate: men, women, children, and families going about the business of living their lives, with all their hopes and fears and dreams. Just so, people who write about the "water crisis," including me, too often fail to consider the actual people whose precious water is poisoned, or scarce, or both — ordinary people who sometimes have to walk great distances to fetch water that is often unclean; people who do it because they must, because without it they cannot survive. We need to remember these people.

The criminal justice system, at sentencing hearings for offenders, allows what it calls Victim Impact Statements from, or about, those most affected. So here is one, about Manya, the senior woman of a household in a small village near the town of Narok in Kenya.

The villagers were members of the Gabbra tribe. I wrote about Manya more than a decade ago, after her family sheltered me from a ferocious riot that had broken out between a group of over-exuberant Maasai warriors and the Kenyan security police. The context of the anecdote was not the riot, however, but water — it described how Manya and her daughters set out, on foot, to fetch water from the nearest well, three kilometres away. Later that evening, the women returned, each with a fifteen-litre pail balanced on her head. They swayed down the trail, singing one of their working songs to pass the hours, as they had done that morning, and as they would do on the days that followed, and as they expected to do, if they thought about it at all, forever. A little later we ate corn mash and fried banana, and sucked on mangos. I declined the water, partly out of politeness and partly out of fear. The well was an old one and had originally been used by fifty families. Now, four times as many drew water from it, and they had to go down further every year. A few months earlier, Manya said, men had descended into the well and had deepened it by two metres or so. The water the buckets brought up was muddy and smelled rank. But they had to drink it, for there was no other. They couldn't clean it, for they had no money. They couldn't boil it, for they had no fuel. They had no choice, and no way to choose.

Inside the one-room family hut were a number of children, most of them listless, lying without moving on their pallets. The youngest, who had not yet been named, was dying. Dysentery, probably, for Manya spent more time than she could afford crooning to the baby and cleaning up its diarrhea, which was considerable. She faced the coming death with fortitude. It wouldn't be the first. There was nothing she could do about it in any case. Nobody seemed to care — even the men of the village didn't seem to care, though she was circumspect about their callousness. What she cared most about were her elder daughters, those who helped her with her tasks, including fetching water. She had hoped, once, that they would get schooling, sometime, somewhere, but it hadn't happened. Nobody cared about that either. Certainly the government didn't care. The Big Men in Nairobi — and the lesser Big Men in Narok — paid the villagers no attention. They didn't figure in politics. They were just peasants.

Manya's daughters were trapped by the family's need for water. She and her daughters were enslaved by this need, though she would never use the word, or even think it. It rippled through the community and beyond. None of the women in her village and in other villages like it — hundreds of them, thousands even — were being schooled. The cost to them and their society is never counted. It's just how it is.

In one way, Manya's is a local problem. If it is ever going to be solved, it will have to be solved locally. You can't help her by being water thrifty in Seattle, or by rationing water in São Paulo, or by desalinating seawater in Perth. But nor is hers just a water problem, to be solved by another well, or a diesel pump, or a solar distiller, though it is that too. Hers is also a political and human development problem, a problem of missed potential — which *is* a global issue. If you multiply Manya up by her millions, as you must to get a true picture of the world, it becomes a human tragedy.

Part One

Assessment

1
The Dismal Arithmetic of Water

That's right, it's come to this,
yes, it's come to this,
and wasn't it a long way down,
ah, wasn't it a strange way down?

— Leonard Cohen, "Dress Rehearsal Rag"

Indeed, it has come to this.

It rained for a while in the second week of November 2014, and then again late in December, but not enough to make a difference. The Brazilian city of São Paulo and its hinterlands, South America's largest metropolis, home to more than twenty million people, the economic engine of one of the prominent BRICS in the edifice of global development, in a country that contains the planet's greatest river and nearly a sixth of all the available fresh water in the world, was within a few weeks of catastrophe, its drinking-water reservoirs down to 2 percent of capacity and at risk of its pumps sucking air and noisome mud, the area's few remaining rivers too polluted to drink and more than ten million people on water rationing, though much too late. In the adjacent commuter city of Itu, police were forced to escort water tankers to prevent hijackings by armed gangs; violent protests by angry citizens vainly demanding the restoration of tap water spread across the region. Politicians were ducking for cover.

How could it *possibly* have come to this?

Through carelessness, cowardly politics, heedless development, burgeoning population, hands-off regulators, deforestation, and climate

15

change. All of them, all at once. And arrogance, that too. There are lessons here for the water world.

For decades, water in São Paulo State has been managed by a water and wastewater utility called Sabesp, a large and until recently rather unruly and poorly managed company that supplied water and sewage services to 363 municipalities across the state. It served more than twenty-seven million customers, 60 percent of the population. The utility was owned by the state government and funded through low-interest loans from its political masters. Half its staff comprised patronage appointments. Its services weren't awful, but they weren't very good either. Leakage rates were higher than 30 percent, much of the region's sewage flowed through open channels and was dumped into rivers after only cursory scrubbing, and its reservoirs were ill-protected from pollution.

This changed somewhat in 1996 when the utility was partially privatized. In that year, a proportion of its shares were listed on the New York Stock Exchange, part of the wave of privatization then cresting across the globe, driven by ideology and the need to raise money without raising taxes. By 2013, the utility was 51 percent owned by the state and 49 percent owned by private investors, and on the face of it things were going well. It *seemed* to be a triumph for a kind-of, sort-of privatization, and thus confounding to the anti-privatization activists. The utility raised much of its funding privately, without recourse to government loans; the political appointees were purged and technical staff hired. As a consequence, the number of employees per thousand connections, a normal way of counting efficiency in the water business, halved in twenty years. The CEO, Dilma Pena, is a woman of formidable energy, as even her competitors concede, and she had taken Sabesp international with signed agreements in Costa Rica, Spain, and even Israel, itself a famously good water manager. Pena crowed that the utility had a policy of "forging partnerships with other sanitation companies, private or public, which want to contribute to the expansion of services." The company website, which carried Pena's comment, almost swooned with admiring self-promotion, its stated aim to "guarantee gains in terms of quality of life, improvement in the health sector and preservation of the environment."

Obviously, this sunny self-image was delusional. What the BBC's Wyre Davies called the "car-crash scenario" of a record dry season at the end of a four-year drought, coupled with the ever-increasing demand for resources, inexcusably caught the utility by surprise.

As it turns out, Sabesp had been expanding services, but not nearly fast enough to keep up with the new people flooding into the region. The sewers are still open channels, and so those rivers that could have been an emergency backup remain unusable. No new reservoirs have been constructed in the last decade, mostly through lack of foresight but also because of red tape in the assessment process. Treatment plants that could have turned wastewater into potable water remained in the file drawers of the company's planners. No desalination facilities were conceived or built. In the run-up to the presidential election in October 2014, no one wanted to suggest either massive new infusions of capital or warn about shortages. Comprehensive water rationing across the region months earlier would probably have helped, but the need was ignored for fear of alarming voters. Sabesp's private investors complain loudly about government interference, which in fact has been minimal; anti-privatization activists, by contrast, blame the profit motive for the lack of long-term infrastructure investment. The leakage rate, despite the hiring of more technicians, remains at the same 30 percent it was decades before. In early November, far too late, Sabesp released maps showing where "water reductions" would occur but still resolutely avoided the words "cuts" and "rationing." (In Itu, as Associated Press reporter Adriana Gomez Licon noted in a dispatch in November 2014, they would have been delighted with rationing, which would at least imply there was water there to ration in the first place.) The Cantareira system, the main water reservoir feeding the region, dropped to just 3.4 percent of its capacity on October 21, 2014, according to Sabesp, and it was still sinking. Pena herself went public to reveal that the city had less than three weeks water left, and maybe only two, unless demand dropped.

Dragged before Congress in late November, the head of the water regulatory agency, ANA, said that after sporadic rain, the city had about two months' supply left. It began tapping into the second of three emergency reservoirs. The last of the three might not even be usable: it is filled with silt, and pumping from it might be difficult or even impossible. ANA's boss, Vicente Andreu, put it this way in response to a question: "I believe that, technically, it would be unviable. But if it doesn't rain, we won't have any alternative but to get water from the mud."[1]

Later in the year, things improved a bit. Rain came, though far too little. And yet the average citizen of São Paulo was still using around 200 litres a day (when it was available), far more than the European average, which is 150. In 2013, one *paulistano* in four actually used more water than the previous year. Water is metered, but only by apartment building, not by individual flat.

Elsewhere in the region, agribusinesses were obliged to transport the soybean and sugarcane harvest by road rather than river because there wasn't enough water to float the barges. The crop yields themselves were lower because of the drought — coffee, sugarcane, oranges, soy, all down by 15 percent or more.

Lost in the hubbub and alarums is the irony that it is those very crops that are a prime cause of the extended drought because of the Amazonian forests that were cut down to plant them. How could that be, when the Amazon is almost two thousand kilometres away from São Paulo? Antonio Nobre, a climate scientist at Brazil's National Space Research Institute, has concluded that climate change and deforestation between them are altering the climate in the São Paulo region by drastically reducing the release of water by rainforest trees — the Amazon rainforest once evaporated about 20 billion tonnes of water every day, and much of that used to drift south. "Humidity that comes from the Amazon in the form of vapor clouds — what we call 'flying rivers' — has dropped dramatically, contributing to this devastating situation we are living today," Nobre told Reuters.[2]

Deforestation in the Amazon basin had slowed in the last decade of the twentieth century and in the early years of the twenty-first. There was no increase in tree cutting until 2012, when the records showed a 29 percent jump in one year — satellites showed that 5,891 square kilometres of forest were transformed into cropland in just twelve months. For the last four years, the flying rivers have failed to fly. And, bizarrely, the Brazilian army has been deployed into the Amazon region on manoeuvres, to "consolidate a doctrine of jungle combat," as an army spokesman put it, to guard against unspecified foreign powers that might want to seize the rainforest for purposes unknown.

Deforestation in São Paulo's immediate hinterlands has made the situation worse. Four-fifths of the forests around the Serra da Cantareira watersheds have been cut down, replaced by farmland and eucalyptus plantations used to make toilet paper, among other things. These are the forests that once acted like giant sponges, soaking up rain and releasing it slowly into watercourses. José Maria Cardoso da Silva, Brazilian adviser to Conservation International, explains it this way: "The forest loss in Cantareira increased erosion, caused the decline in water quality, and changed seasonal water flows, reducing the resilience of the entire system against climatic extreme events."[3] Water once captured by the forest and funnelled into reservoirs caused mudslides instead.

But the cutting continues. The government of newly re-elected Dilma Rousseff has made no effort to scale it back. Indeed, it has encouraged its expansion; the building of hydroelectric dams on the Amazon's tributaries continues apace, encouraging further development and the cutting down of ever more trees.

I dwell on São Paulo, not because it is the direst situation facing the water world, but because it encapsulates the debate about what has gone wrong with the planet's water and the arguments about how to fix it. Folded into the discussion are some of the hot-button issues I raised at the start of this book: whether privatization of water systems can help or whether it makes everything worse; whether bulk transport of water is a fix or a danger; whether the multiple water crises are a function of inevitable natural constraints or a failure of management (or both); and whether, indeed, there is a single crisis that needs to be tackled globally, or thousands of local and regional crises that would have to be tackled where they occur. And, of course, climate change.

Any cursory survey of current writing on the state of the world's water will turn up multiple iterations of the catchphrase "The dismal arithmetic of water." True, the World Commission on Water (subsequently widely quoted in World Bank documents) called it "the gloomy arithmetic of water," occasionally written as "the grim arithmetic of water," but they all mean the same thing. Well, "dismal" is a nice word, softer than "appalling" and easier on the tongue than "execrable." But what it means is simply this: many parts of the world are short of safe, clean water, and the shortages are getting worse, with no end in sight.

The way out, according to the World Bank and its fellow travellers, and also to the Chinese, is massive investment in infrastructure, the building of huge new reservoirs, and other technological and engineering solutions costing billions, and even trillions, of dollars. This is the so-called "hard path," the route of the technocrat. Opposed to this is the "soft path," the road to increased water efficiency and reduced water demand. Both sides in this debate think the other wrong-headed.

The main question is the degree to which water's problems are a factor of natural constraints — in which case, the limits are real and fixable only with substantial effort and cost — and how much is due to poor management — in which case, the remedy is obvious, if elusive. Good management, as we can plainly see from the state of the political world, is something of a hard premium in our times. Both paths will be explored in subsequent chapters.

The American water guru Peter Gleick, who runs his own water think tank called the Pacific Institute, has suggested that we contemplate the notion of peak water, a concept that mimics the much-debated and much-contested notion of peak oil, in which production peaks and subsequently declines, thereafter falling short of total consumption.

With water, Gleick goes further, suggesting three peaks, not one: peak *renewable* water ("where flow constraints limit total water availability over time"), peak *non-renewable* water (similar to peak oil, "in which water consumption rates substantially exceed natural recharge rates and where over-pumping or contamination leads to a peak of production followed by a decline"), and peak *ecological water* (defined as the point beyond which the total costs of ecological disruptions and damages exceed the total value provided by human use of that water).

In his measured fashion, Gleick says that "despite uncertainties in quantifying many of these costs and benefits in consistent ways, more and more watersheds appear to have already passed the point of peak water."[4]

Looked at another way, there are two dismal metrics, two overlapping water crises: the first is dirty water, the contamination of water supplies, even in areas where water itself is relatively abundant; the second is where water, dirty or clean, is in short (sometimes critically short) supply. They overlap because some areas that are water stressed are similarly bedevilled by pollution, making solutions incrementally more complicated. Both are further complicated by local, regional, national, and sometimes transnational tensions, sometimes arising from water directly, in other cases where water is a surrogate for greater conflicts.

Water quality is poor everywhere — hardly any rivers are now fit to drink in their "wild" state — but in the rich countries this can be, and is being, overcome. True, attitudes to water quality need some adjustment (the "new normal" of boil-water advisories, no-swimming advisories, and so on), and the water itself too often needs expensive infrastructure to purify supply, but these are now pretty much universal. In poor countries, there is clear evidence that where population increases and economies do not prosper, matters are getting worse, not better, despite decades of work by the aid agencies. A decade ago, Peter Gleick suggested that if water and sanitation services did not radically improve, as many as 135 million people would die from water-related diseases over the next twenty years, about the same number as those killed in all the wars in all of human history. That prediction sounds extravagant, but it actually came true — that many people did die of preventable water-borne diseases, a decade ahead of "schedule." Matters have improved since then, but there are *still* a quarter billion new cases of water-borne diseases every year, killing the population equivalent of Canada every three years or so. Somewhere in the world, a child *still* dies every six to eight seconds from drinking contaminated water — that adds up to three to five million a year. Malaria, despite the existence of cheap and effective prophylactics, kills more than a million people a year. Bilhfarzia and dengue fever are spreading, not retreating. New tropical diseases such as chikungunya have made their way from Africa to the Caribbean just in the last few years. Cholera and dysentery risk becoming endemic. Even in Europe, somewhere around forty children die each day from drinking unsafe water.

As noted at the start of this book, the population of our little planet has tripled in the last century but water use has sextupled, about three-quarters of it used just to grow the food to feed the world. Some of the consequences are obvious: dozens of rivers no longer reach the sea. China's Yellow River, or Huang He, now makes it to the ocean only in flood times; America's Colorado is a mere drainage ditch by the time it gets to Mexico; you can often walk dry-footed across the Rio Grande; the Jordan River has become a pathetic trickle, and so on. Dismal news indeed. At least half the world's wetlands, once nature's flood-control mechanism, have disappeared in the last century. Forests, "nature's sponges," are disappearing everywhere. The Florida Everglades, usually counted as a conservation success story, are less than half their original size. Water supplies in the Nile Valley are in peril. A fifth of the planet's freshwater fish that existed a century ago are already extinct. Groundwater tables, the aquifers on which so much of irrigation depends, are becoming seriously overdrafted in many parts of the world. Water use will likely increase by 50 percent in the next thirty years — if the water can be found. Somewhere between a tenth and a fifth of the world's food supply is being irrigated by over-mined aquifers.

The US national security apparatus, alarmed by the potential for water conflicts and for floods of ecological refugees, has calculated that in the next thirty years around four billion people — somewhere between a third and half of the global population — will live under conditions of severe water stress, which means living without the water to meet basic needs. Shortages will be particularly severe in Africa, the Middle East and Arabia, and South Asia, though many parts of the Americas will be similarly afflicted. As always, the poorest countries and their poorest people will be most directly damaged.

As the authors of the World Bank's 2003 Water Sector Strategy paper put it, this gloomy arithmetic of supply is mirrored by an equally gloomy arithmetic of costs. The easy and cheap "hard" options for increasing the water supply have already been tapped. Most of the easy rivers have been dammed. A decade ago, the World Bank declared that necessary water infrastructure investments would have to increase from around $50 billion a year to more than $180 billion, just to keep from falling behind. The number is now far higher.

To manage the water world properly, you first need to know how much water there is, and where it is to be found, and then, having calculated that, it is necessary to know how much of that water is available — usable on an ongoing, sustainable basis. None of this is easy, even after decades of meticulous accumulation of data. Monitors mostly exist where they are less needed — in Europe and North America — and are spotty and unreliable in the rest of the world. Also, it is still not known how much fresh water is stored in permafrost and glaciers. There are some guesses, but even the guessers themselves admit the computations on which they are based are crude.

Water is no more than 0.2 percent of planetary matter, but despite this, most of the earth's surface is water, albeit mostly salty or locked into ice caps. The notion that we would ever "run out" of water is ludicrous. There is more than we could ever use. But usable water is another thing. *That* we can run out of.

The renewability factor is critical. As an example, calculate how much water the North American Great Lakes "really" hold for human use. Lake Superior is not the largest of earth's lakes (Baikal in Russia and Tanganyika in Africa contain more water), but the Great Lakes system does contain somewhere around 27 percent of the easily reachable surface fresh water on the planet. These lakes replenish slowly — if we were to take out a percentage point or so more water than that which goes in, the water levels would drop — and keep on dropping. That's why there is no conflict between the statements "The Great Lakes contain a quarter of the world's fresh water" and "Canada has only 6 percent of the world's renewable water." Canada has, in fact, almost identical renewable resources as China, in more or less the same scale landmass. Except China, of course, has thirty times the population, so the per capita availability of water in China and Canada are utterly different — one place is stressed, the other is not.

Thus, it is the flow of water, not the total amount, that gives us our livable quotient. And this flow is highly variable, affected by weather, climate, and a host of other factors. So here are some necessary numbers:

If there are almost 1.5 billion cubic kilometres of water on the planet, as is plausible, a mere 35 million of those are fresh water and most of that relatively trivial amount is not accessible to us — about 24 million cubic kilometres are locked into the Antarctic and Greenland glaciers and permafrost. The rivers, lakes, wetlands, and aquifers from which we derive usable water amount to somewhere around eleven million cubic kilometres. A *Science* paper by Taikan Oki and Shinjiro Kanae put the global total of available (that is, sustainably usable by humans) water at 45,500 cubic kilometres.[5] Igor Shiklomanov, of the State Hydrological Institute in St. Petersburg, Russia, calculated the figure at 42,000 cubic kilometres. My earlier book on water, a decade and a half ago, suggested a more conservative 34,000 cubic kilometres, a number based on plausible data sets from multiple national agencies. I'd stay with Shiklomanov on this: his institute is constantly updating the data as they are developed, and he brings a properly skeptical mind to the computations. We're actually withdrawing about half of this amount each year. So there is a lot left over.

Is it (will it be) enough?

To begin to assess this, we must look first to the obvious fact that the geographical distribution of sustainable water is unequal — a climatic truism. To a degree, this is our fault: humans have colonized places where there is not enough water to support them (Los Angeles comes to mind). Thus, there are places on earth that have too much water, and places that haven't nearly enough — that water is often in the wrong places at the wrong times and in the wrong amounts has long been a hydrological cliché. Further, climate change suggests that droughts (and floods) are increasing, which in turn means that stress is appearing in places that had never imagined it before. At the same time, it is also obvious that water is heavy and not easily transported. The Amazon basin accounts for almost 20 percent of the world's run-off, but you can't feasibly shift that to countries that are short — hell, you can't even shift it to São Paulo, in the same country. There is no practical way of solving one region's shortages with another's surplus. In this sense, a suggestion by the International Rivers network that "just one percent of current [global] withdrawals" would give everyone who needed it forty litres a day is disingenuous: How are you going to get the water where it is needed?[6]

A second point is the equally obvious fact that the easiest aquifers have been tapped, the easiest rivers dammed, and all the easiest sources already discovered, and that we have contaminated more water than is ecologically sensible. This means that solutions become more difficult. Not impossible, just more difficult.

And finally, we look at the per capita use of water around the world. It has been increasing steadily, for a variety of reasons. Population growth increases total usage, not per capita usage — more people means more water use, even if the users are poor. But if they cease to be poor, they use even more — they use more water for sanitation, and for growing food (especially if they switch to eating more meat), and for multiple other purposes, including more manufacturing. The numbers illustrate the point: on a global scale, per capita water *availability* has decreased from 12,000 to 7,600 cubic metres per person, that is, by more than a third. Africa has seen the greatest change — per capita available water decreased there by 2.8 times. But the numbers are not trivial elsewhere either: Asia was next, with per capita availability decreasing 2.0 times; South America, 1.7 times; and Europe, 1.26.[7] Total usage has gone up, sometimes dramatically: if consumption is expressed in cubic kilometres per year, usage went up in the twentieth century from 37 to 463 in Europe, 69 to 705 in North America, 40 to 235 in Africa, and 414 to 2,357 in Asia.

This doesn't mean that the situation is beyond control, or salvation. So far, only North Africa and the Arabian Peninsula are in overall water deficit, extracting more water from their reservoirs than nature is replenishing (both regions are using around 130 percent of available supply), though the use in parts of Asia is approaching 80 percent of available supply. Of course, these national and regional figures mask local and river-basin critical areas — the fact that the United States is in reasonable shape doesn't mean the Ogallala aquifer isn't being depleted.

To some degree, "virtual water," sometimes known as embodied water, can alleviate stress in parched regions. Half a kilo of coffee beans, for example, has a "water footprint" of more than eight litres — that is, it takes eight litres of water to produce those beans, yet the beans are much lighter than water and can be easily transported — much more easily than the water that produced them. So Israel, say, can "conserve" water (that is, not have to use it up by irrigating coffee plantations) by buying coffee grown in places that are relatively water-rich. If managed right, both sides benefit.

Despite appearances, the unequal distribution of water, the fact that the remaining supplies will be harder to get than heretofore, and the increase in per capita consumption are amenable to solution. The solutions won't always be easy, and will often be expensive, but they are achievable.

2
The State of Our Groundwater

Every spring I take the heavy concrete lid off our garden well and peer into it, to see where the water is. Then I usually check again in late summer. The latest year had a wet spring that turned into a for-est-fire-danger summer, but in our well the water had hardly dropped at all. The well is only seven metres deep, but it has five metres of water in it. The water remains clear throughout the year and tastes clean, if a little flinty — we could probably hire a water sommelier and sell the stuff in bottles for twice the price of high-test gasoline. But, of course, if we did that, and set up a bottling line nearby, and started to pump out the water in bulk (or "draw it down," in the jar-gon of hydrologists), the water table — that's the surface of the water, two metres down — would start to drop. Pump the water in sufficient quantities, and after a while the pumps would start to suck air — no more Eagle Head Premium mineral water.

Of course, that's just fantasy. What we're drawing from our well, and thus to the aquifer that underlies it, is nowhere near enough to drop the water table. We're just a small family, a cat, and a vegetable garden, and we don't come close to even matching the natural inflows. But that's not true elsewhere, and on scales much more important than our little household. All across the planet's surface, powerful diesel pumps (or, in many cases, multitudes of lesser pumps) are drawing groundwater to the surface for irrigation. Much of the world's food supply depends on this water. If enough of those pumps start to suck air, the consequences will be altogether more dire.

But it needn't be so.

Numbers compiled in 2012 suggest that better than 70 percent of all human water use (and up to 90 percent of what is called consumptive water use — that is, water that cannot be reused for other purposes) is for growing food. Much of agriculture still uses rainwater, but groundwater irrigates somewhere around a fifth of all croplands and helps produce as much as 40 percent of the world's food. Groundwater use for irrigation is also increasing, both in absolute terms and in percentage use, leading at times to what the authors of a comprehensive study call, rather circumspectly, "concentrations of users exploiting groundwater storage at rates above groundwater recharge."[1]

The amount of groundwater used for irrigation varies greatly by region. Africa, which is generally too dry to rely on rain, uses it for three-quarters of food growing; Europe, by contrast, draws on groundwater for only a fifth of its food-growing area, and that mostly in the south. In global terms, the biggest groundwater users are India, which irrigates thirty-nine million hectares; China, nineteen million hectares; and the United States, seventeen million hectares. Nearly 90 percent of Indian irrigation comes from groundwater.

Groundwater is, of course, used for many purposes other than irrigation. More than 70 percent of the water used for all purposes in the EU comes from groundwater. Aquifers, where groundwater is stored, are often the major source of water (and often the sole source) in some arid or semi-desert countries — 100 percent in Saudi Arabia and Malta, 95 percent in Tunisia, 75 percent in Morocco.

Groundwater is important on short timescales in another way: since it responds much more slowly than surface water to meteorological conditions, such as the changing climate, it can provide a natural buffer against drought.

Because groundwater makes up almost the entire volume of the earth's usable fresh water — almost 96 percent — the accumulated volumes in rivers and lakes are trivial by comparison. Aquifers large and small are everywhere. Some of them are very deep — up to three kilometres below ground. And some are very old — aquifers deep below Canada, South Africa, and Scandinavia contain water estimated to have been there for a billion years.[2] Many of them are too small to notice or to name; they provide water for individual farms, or families, or small villages. Others are much larger, and the most famous are immense. Substantial aquifers have been found in areas now arid or desert, such as the Sahara, some of the ranch lands in South America, and under the High Plains in the United States.

There are a number of claimants to the title of world's largest aquifer. They are to be taken with large pinches of salt because some of them are measured by the land area under which the aquifer lies, while others are measured by the supposed volume of water they contain — as a hydrologist once told me, "Apples and oranges ain't in it." A fair consensus is that the North African combo (Nubia, Murzuq, and the rest) that jointly covers more than two million kilometres is at least the world's largest fossil-water aquifer. The other claimants are the High Plains aquifer in the southwestern United States, and the Australian Great Artesian basin (1.7 million square kilometres, claimed as the largest artesian aquifer in the world), and South America's Guarani.[3]

UNESCO and the International Hydrological Programme have catalogued the world's many transboundary, or international, aquifers. As of 2011, they had mapped 273 such boundary-crossers: 68 in the Americas, 38 in Africa, 90 in Europe, and so far 12 in Asia, where the survey has yet to be completed. As a UN General Assembly resolution pointed out, some of the largest aquifers remain unexploited, such as South America's Guarani and North Africa's Nubian.

A survey of known aquifers shows some surprising gaps in our knowledge. Canada, for example, is water-rich and has not found it critical to map its groundwater — though a third of the population depends completely on groundwater for drinking. It wasn't until 2009 that a concerted effort was made to do a proper inventory, an effort "to establish a conceptual framework of national, regional and watershed-scale groundwater flow system."[4] This mapping is ongoing.

The United States has a better grasp of underground water, partly through the busyness of the US Army Corps of Engineers, and partly through the federal and state geological services. They range from the immense (High Plains/Ogallala) to the small-but-important (the curiously named Mahomet aquifer underlying parts of Illinois). Some are better known because food growers or urban dwellers depend on them, such as the Santa Clara Valley aquifer, the Texan Permian sea (mostly known for its oil, not water), the Magothy underlying Long Island, the Biscayne aquifer of South Florida (one of the shallowest aquifers, sometimes hard to distinguish from surface lakes and streams), and the Snake River aquifer of Idaho (everyone's ideal, yielding plenty of water for agriculture but hard to diminish, recharged as it is from snowmelt).

Mexico has an arid and well-populated north (more than three-quarters of the country's population and almost 90 percent of its industrial output). The south is poorer, less developed, and a good deal wetter. In total, the country has a well-mapped groundwater inventory: more than 650 aquifers have been identified, with the most contentious, and abused, one under Mexico City itself. Three watersheds and more than a dozen aquifers are shared with the United States. Agriculture uses three-quarters of the groundwater supply, but the majority of cities and towns in Mexico also rely on groundwater for their drinking and sanitation supply. If you look at the gross numbers, the country seems relatively well-off: groundwater recharge is estimated at nearly eighty cubic kilometres a year, while withdrawals amount to less than thirty cubic kilometres. But as elsewhere, these numbers can hide real problems. The north is generally in overdraft, and getting worse. The south, with plenty of water to spare, nevertheless has problems supplying clean water — many of the aquifers are heavily polluted, sometimes to the extent of becoming unusable without expensive treatment.

As of 2015, southeastern Brazil notwithstanding, no countries in South America, not even the most arid, are undergoing a real water-supply crisis. But the situation is deteriorating. For one thing, water-use patterns have been shifting, and many more countries than before are increasingly relying on groundwater. Pretty much everywhere, groundwater use is rising as a proportion of the whole supply, mostly because it is less polluted, but also because of the rising costs of maintaining safe water in above-ground watercourses and storage. This increase in groundwater use includes, counterintuitively, the Amazon sedimentary basin, which you would think has surface water to spare, but Amazonian cities such as Manaus, Santarem, and Belém all rely on wells for their water supply. Many South American aquifers are shared between two or more countries. So far, there have been no serious conflicts over water.

The two most interesting aquifers in South America are the Guarani, explored in more detail later in this chapter, and the Hamza, unique among the world's aquifers for being both river and aquifer — an aquifer, flowing steadily west to east four thousand metres beneath the Amazon and emptying deep below the surface in the Atlantic Ocean. The Hamza, named after its discoverer, geophysicist Valiya Hamza, has only a fraction of the Amazon's flow, perhaps 4 percent, but it is moving quickly — maybe three thousand cubic metres pours into the sea every second.

In the Middle East, the most important aquifers, not surprisingly, are those shared by Israel and Palestine. Israel and the West Bank rely on the Mountain aquifer, usually referred to in Israel as the Yarkon-Taninim aquifer. The aquifer underlies the Dead Sea and is vulnerable to saline intrusions. It is unequally split; Israel uses seventeen times as much water from it as does Palestine. The situation in Gaza is dire. The small Coastal aquifer, really the only source of water in the strip, receives somewhere around 50 million cubic metres of recharge every year, mostly in rain runoff from the Hebron Hills, but Gaza's people (and nearby Israeli farmers) are drawing out 160 million cubic metres. As the water table drops, seawater seeps in from the Mediterranean, and this problem is compounded by the vast amounts of raw sewage dumped into the sea, some of which is making its way back through the seepage.

Europe is reasonably well supplied with groundwater. In fact, Europe is as well-off as anywhere on earth, except possibly New Zealand. Even so, ten countries in the EU, most of them in eastern Europe, have reported groundwater overexploitation, and there are no doubt more that have failed to come clean to Brussels. The reasons are the usual ones: increasing amounts of water taken for private and industrial uses, mining, irrigation, and the rest. Drought too — there have been increasing dry periods in much of eastern and southern Europe. So far, no countries except Malta have reached critical status, but there have still been consequences: wetlands are disappearing, and saltwater intrusions, sometimes from the sea and sometimes from heavily (and naturally) mineralized deep aquifers, have been detected, necessitating costly remedial programs. Burgeoning populations and the push for economic growth have also fuelled this decline.

One of the pleasant surprises in the global water inventory is Africa, especially north-central and northeastern Africa, one of the driest places on the planet, where farmers and pastoralists have struggled for centuries to coax water from what wells they could manage.

Water in serious quantities was first discovered under the desert in the 1970s, in the Al Kufrah region of Libya, in the southeast quadrant where it converges with Egypt, Sudan, and Chad. It was an accident — the prospectors dispatched into the interior by Libya's late unlamented dictator, Moammar Gadhafi, were looking for oil, not water. What they found instead of black gold was blue gold. Subsequent test wells drilled in the region told the same story: underlying the desert, in the saturated sandstone and in the interstices of the underlying shield, was an immense reservoir of water which came to be called the Nubian Sandstone aquifer. It is thought to contain around 375,000 cubic kilometres of water — the equivalent of 3,750 years of Nile River flow, a timespan long enough even by Egyptian standards. Yields from wells drilled into them can be astonishingly high, up to 700 cubic metres of water per hour.

In 2012, a team of research scientists from the British Geological Survey (BGS), in collaboration with University College London, mapped in some detail the amount and potential yield of North Africa ground-water generally. One of the scientists, Helen Bonsor, had this to say: "Where there is the greatest groundwater storage is ... in the large sedimentary basins, in Libya, Algeria, and Chad. The amount of storage in those areas is equivalent to 75 metres thickness of water across that area." Then, briefly lapsing from her careful scientist's vocabulary, she said, "It's a huge amount — huge."[5]

Similar satellite mapping techniques have found aquifers under Namibia, under the Volta basin (Burkina Faso and Ghana), under the Sokoto-Iullemeden basin in northwest Nigeria, under the Kalahari Desert, under South Africa's Great Karoo, and more.

The BGS study estimated that total groundwater storage in Africa was somewhere around 0.66 million cubic kilometres, more than a hundred times larger than annual renewable freshwater resources. Like any other water source, over-abstraction would ultimately exhaust the resource, but for many African countries (this from the formal report), "appropriately sited and constructed boreholes can support hand-pump abstraction, and contain sufficient storage to sustain abstraction through inter-annual variations in recharge."

Then, in 2013, two more massive aquifers were found in north Kenya, currently the home mostly of nomadic herders who are particularly vulnerable to lack of rain. The aquifers were found to underlie the Turkana region and the Lotikipi basin, and were discovered through analysis of satellite and radar images that mapped ancient river courses and trace moisture. Turkana is one of the hottest, driest, poorest regions of eastern Africa, suffering in recent years from armed incursions of bandits from the failed state of Somalia. Test drilling yielded estimates of 250 billion cubic metres of water (to compare, Kenya as a whole uses some 3 billion cubic metres a year). Kenya's environment minister, Judi Wakhungu, announced the discovery at a UNESCO meeting in Paris. In the clichéd language of such meetings, she put it this way: "This newly found wealth of water opens a door to a more prosperous future for the people of Turkana and the nation as a whole. We must now work to further explore these resources responsibly and safeguard them for future generations." And she hastily added, reacting to early rumblings from communities and NGOs in the Turkana region, "The first priority, of course, is to supply water to the people of the area, who have always been water insecure."

What this means is that a number of countries thought to be water-scarce are in fact relatively water-rich, and therefore more able to cope with at least some of the changes brought about by changing climate. This is not to say they can't be depleted — large-scale pumping, as proposed by Chinese engineers who have grabbed up large swaths of land with a view to exporting food back home — would deplete even the most robust aquifers in time. But, as Bonsor told the BBC, "Our work shows that with careful exploring and construction, there is sufficient groundwater under Africa to support low yielding water supplies for drinking and community irrigation. Even in the lowest storage aquifers in semi arid areas with currently very little rainfall, ground water is indicated to have a residence time in the ground of [only] 20 to 70 years. So at present extraction rates for drinking and small scale irrigation for agriculture groundwater will provide and will continue to provide a buffer to climate variability."[6]

Australia, the driest landmass of them all, is reliant on groundwater for most of its water security. Groundwater underlies almost a quarter of the Australian landmass, and is very deep in some places — as much as three thousand metres. This is really the only reliable water source for the interior. It may — no one knows for sure — contain somewhere around sixty-five thousand cubic kilometres of usable water. For the rest, there are numerous coastal aquifers, including several around Perth, but most of them are threatened and Perth is looking to desalination for its future use.

The State of the Aquifers
In arid regions on every continent, water tables are dropping because of over-pumping. In the Punjab and Bangladesh, the table is dropping a metre a year, complicated by the lamentable fact that deeper wells have to punch through a layer of natural arsenic, making the water

quality poor and even dangerous. A study by Matthew Rodell and others published in *Nature* in 2009 ("Satellite-Based Estimates of Groundwater Depletion in India") reported that in north India the water table had been dropping at a steady four centimetres a year for decades. In the study period, from 2002 to 2008, groundwater losses amounted to 109 cubic kilometres, double the capacity of India's largest surface-water reservoir, while rainfall and other sources remained normal — proof that human overuse caused the losses. In the North China Plain to the west and north of Beijing, it is dropping even faster, and water is being diverted from agriculture to industry, a consequence of explosive growth and reckless development (Shijiazhuang, a small- to medium-size city of two million inhabitants, has been advertising "waterfront property" on lakes filled with pumped groundwater). Where once farmers could strike water at two metres, now they struggle at twenty, and sometimes much more — virtually all the natural wetlands in the north have vanished.[7] In Thailand, Bangkok is pumping so much water from its aquifer that the city is sinking by about ten centimetres a year and will be below sea level by 2050. Nine Indonesian cities are dropping by almost thirty centimetres a year, due entirely to groundwater over-pumping. In North America, much of High Plains agriculture and some city water supplies are dependent on the Ogallala aquifer, and it is no great secret that the aquifer levels have been dropping steadily as its water is mined.

In 2012, a study by McGill University engineers found that almost a quarter of the world's people now live above a threatened aquifer. The informative website Circle of Blue quotes one of them, Tom Gleeson, as saying, "A lot of agriculture depends on the unsustainable use of groundwater. That can and will affect agricultural production."

In the last year or so, scientists have discovered an entirely unsuspected, but very large, reservoir of underground — or rather, underwater — fresh water. Significant amounts of fresh or "lightly saline" or "brackish" water has been detected underneath the continental shelves, laid down under land that was not submerged by the sea until the ice age ended, some twenty thousand years ago. The reservoirs — estimated to be one hundred times all the water extracted from all the aquifers since measurement began, perhaps half a million cubic kilometres — could, according to the Australian scientists who discovered it, "be economically processed into potable water."[8]

This discovery can be added to the relatively new notion of aquifer storage and recharge, or ASR. Essentially this means capturing storm runoff, in newly created artificial wetlands, for injection into depleting aquifers. A pioneering project on the outskirts of the Australian city of Adelaide has a capacity of fourteen million litres, with the resulting water being used mainly for industry in winter and for watering parks, golf courses, and playgrounds in summer. Perth too is injecting treated wastewater into its aquifers, especially the Gnangara, and is "storing" several billion litres a year. Some American jurisdictions have cautiously followed suit. Oregon is storing somewhere around three and a half million cubic metres of flood waters a year in this way, even generating modest amounts of electricity through the pressure of water flowing back underground. The Texas cities of El Paso and San Antonio are doing the same thing, so far on a more modest scale — probably because they have less stormwater runoff in the first place.

At the same time, the UN is attempting to codify and rationalize international management of groundwater. The draft articles for a law on transboundary aquifers was adopted unanimously in 2008, but no one expects its implementation to go as smoothly. As a hedge, Article 3 of the draft recognizes that "each aquifer State has sovereignty over the portion of a transboundary aquifer or aquifer system located within its territory. It shall exercise its sovereignty in accordance with international law and the present draft articles." And then subclause (d) of Article 4 says this: "[Aquifer states] shall not utilize a recharging transboundary aquifer or aquifer system at a level that would prevent continuance of its effective functioning."

Well, they *should* not. But *shall*?

Aquifer Case Study: The Ogallala

It has been clear for years that the real enemy of the Ogallala, and the larger High Plains aquifer of which it is a part, is not the Keystone XL pipeline or the drillers hovering around hoping for a fracking permit, or even pollution. The real enemy is the farmers that use its water. Too many of them, using too much water, growing too many water-hungry crops like corn, for too many of us — too many consumers all round.

Like many other sprawling aquifers, the High Plains aquifer is not a single pool of saturated rock but a regional system comprising a grab bag of smaller units geologically and hydrologically connected, underlying eight US states, from Wyoming and South Dakota in the north to the Texas Panhandle in the south. It can be shallow in places, even reaching the surface at Lake Scott State Park in Kansas, but in others it can be very deep, dozens of metres below the surface. Its water-bearing strata can be as shallow as fifteen metres, and as thick as ninety.

Recharge is slow, partly because rainfall is scant throughout, but also because what rain does fall comes in the growing season, and the plants absorb much of it. The long-term average recharge in many places can be less than three centimetres a year, ranging up to the still-not-abundant fifteen centimetres year in places like central Kansas.

The grim arithmetic of the Ogallala is this: water drawn out is measured in feet. Water put in is measured in inches. It is the very definition of a slow-motion train wreck.

In some areas, the available water has declined only slowly and, with conservation, its decline can be arrested. In places, declines of 30.5 metres or more represent only half the available water, with plenty remaining. In others, declines of 60 percent are common — more than 1.2 metres a year during the current and recent drought. In some places, there is no water left at all. Huge swathes of Texas have now been withdrawn from farming altogether.

The early settlers in the High Plains simply put up with what they found. They had no choice. Oh, sometimes they invested in crank nostrums like no-water seeds, and charlatans and mountebanks were common, inducing rain from cloudless skies for a decent fee, but generally they prayed for rain and used it thriftily when it came, and tolerated the dry years, adjusting their farming accordingly. What technology they came to employ was eminently sustainable: primitive drilling rigs would sink a borehole into the earth until water was struck. On the surface they erected an iron lattice tower with a clanking windmill atop it, and the water was pumped into a cement or corrugated iron tank for distribution either to cattle or to crops. There were few enough farmers and they pumped the water slowly enough — a dozen litres or so a minute was good — that the water table stayed stable for decades. Those iron windmills were a feature of the plains landscape for generations; some of them can still be seen, and a few are still in use.

About six decades ago, a generation after the Dust Bowl, came the arrival of efficient diesel pumps. Fuel costs were so low that farmers could afford to run the pumps on a twenty-four-hour cycle, pumping up prodigious amounts of water, sometimes close to four thousand litres a minute. This changed everything. Farming was no longer a hardscrabble affair teetering on the edge of subsistence, but a profitable business, growing crops to send across the nation and thereafter for export. The efficient irrigation became incrementally more so with the invention in the late 1950s of centre-pivot irrigation systems, mile-wide mobile sprinkler systems that give the landscape those emerald-green crop circles so evident from high-flying aircraft. Now farmers could grow corn, four hundred hectares of it (one hundred thousand hectares in 1950; over eight hundred thousand now), which in turn led to a booming rural economy. As much as 90 percent of the water pumped from the Ogallala is for farm use. The aquifer has sustained not just the Kansas and Nebraska corn and wheat belts but much of Texas beef production too.

But the pumps that had been pulling close to four thousand litres a minute began to pull two thousand, then a thousand, then, in places, a trickle. The hardest hit is a seven-hundred-kilometre rectangle from Lubbock, Texas, to the Kansas–Nebraska border.

The Texan 2012 State Water Plan estimated that the state would face a shortfall of ten trillion litres a year by 2060. The number was widely derided — the plan was said to overestimate future farming demand (dwindling water has already forced corn and cotton farmers into major retrenchments) and to underestimate conservation measures. The revised estimate? A shortfall of "only" four trillion litres a year. The appropriately named Ken Rainwater, a water resources specialist at Texas Tech University in Lubbock, put it simply: "The plan can't make there be more water in the aquifer.... Residents may hope that the current level of farming can continue through 2060, but that's just not going to happen."[9] If depleted entirely and not used again, the Ogallala would take some six thousand years to fill naturally. In the current climate.

Farming will survive. Farming with rainwater is possible, though much harder than with groundwater and with much slighter yields. Farmers will switch crops — some are abandoning corn for, say, dairy cows, which, anti-meat propaganda notwithstanding, use far less water. Several thousand head can survive on water irrigated by a single crop circle.[10] Water will be used more efficiently, resulting in less used and limited acreage irrigated. This is as it should be. It should have been done a long time ago. But at least it is a beginning.

As an example, the North Plains Groundwater Conservation District in 2005 set an annual pumping limit of seven million litres per hectare; in 2012 it was reduced to five million litres, and further reductions are contemplated. Another is from Sheridan County in northwest Kansas, population 2,556 at last census, where the locals have, without prompting, agreed to across-the-board water cuts of 20 percent. Nobody ordered these cuts. In fact, the mess of state and local laws make a mandated remedy impossible, especially in the current dysfunctional political atmosphere. The cuts were agreed to in community meetings called by the farmers themselves. They agreed to short-term pain for an uncertain but slightly more confident future. They want to preserve the aquifer on which they depend.

There's a thought.

Aquifer Case Study: Texcoco

The fate of the Texcoco aquifer (formerly known as Lake Texcoco) is a classic case of human folly — in this case, the building of a city in a place that made no environmental sense. In this, it is analogous to dozens of similar human follies in other countries — cities in the desert where there is no natural water (Las Vegas and a dozen other southwestern US cities); housing developments on barrier islands in hurricane country (the Carolinas, all the way to Massachusetts); the seeding of wheat fields on the edge of the Gobi Desert. It is not anti-Mexican to point out that Mexico City should have been built almost anywhere else.

Lake Texcoco is, or rather was, a natural lake, one of a series in the Valley of Mexico, a high plateau with an average elevation of around twenty-two hundred metres. It was the largest of five interconnected lakes, and the lowest. This meant that in high-water times, the other four spilled over into Texcoco. The other thing to remember is that the Valley of Mexico is a closed basin, with no natural exits, a recipe for flooding.

The Aztecs built a city on an island in the lake, prudently above high water. They built it there because it was easily defended against raids, with a kind of natural moat. The city was small enough to be easily contained on the island, and flooding in the surrounding lands was moderated by an extensive systems of dikes, canals, and land bridges, or causeways. It worked just fine for several centuries.

After the conquest, the Spanish built what came to be known as Mexico City on the newly flattened ruins of the Aztec fortress-city. But it grew too big, spilled off the island, and was periodically inundated. In 1607, after a particularly nasty flood, the authorities constructed a drain to lower the lake levels. It didn't work — the water stayed in the basin, and washed back in. A decade or so later, another flood inundated the city for more than five days. There was some thought of relocating the city elsewhere, but the folks at head office (Spain) vetoed the move and suggested a tunnel. That didn't work either — the city was by now mostly beneath the water table. And so it went, flooding now and then, with resigned periods of rebuilding and drying out. Still no one moved away, and the city kept growing.

It wasn't until the twentieth century that a solution of sorts was found. The government of the time constructed an extensive network of deep tunnels, some of them as far down as 250 metres, to punch through the basin wall and so out into the surrounding countryside. It worked too, but had unforeseen though entirely predictable consequences. Most of the Valley of Mexico is now arid and can be farmed only with difficulty. Mexico City (population 10.8 million in 2010, but 21.2 million in Greater Mexico City, making it one of the largest conurbations in the world) has to bring in its water for drinking and sanitation, via a network of pipes, from several hundred kilometres away, including from the Bravo Valley basin and four "well fields" in the surrounding countryside. Water is also brought in from the Cutzamala River to the Southwest, through tunnels and aqueducts, but it has to be lifted more than a thousand metres before it can flow down to the city and the system is operating at less than half its designed capacity. The depletion of the local water table has meant the city is sinking, in most places only a few centimetres a year, but parts of the city have dropped twenty metres in fifty years. This has cause an estimated 35 percent of the water to leak away because of ruptured (and rusting) pipes. The city is therefore chronically short of water, and the quality of the water it does have is poor.

Money could probably fix the system — large mountains of money. It is hard to see any other remedy.

Aquifer Case Study: The Guarani

Until a few years ago, no one outside the specialized world of hydrology had ever heard of the Guarani aquifer, this despite the fact that it underlies more than a million square kilometres of Uruguay, Argentina, Paraguay, and Brazil, including a somewhat unsavoury and densely forested region known as the Triple Border. It was not very well known even in the four countries concerned. Even the locals who drew on its water didn't really "know" it in any meaningful sense.

That has now changed, and not entirely because of the wild conspiracy theories that surround and threaten to drown it. It has changed because it *is* huge, it does contain significant amounts of so far still-pristine fresh water, and it *is* transboundary in a region where boundary irritations are commonplace. The aquifer even made a somewhat disguised appearance in the James Bond movie *Quantum of Solace*, though the movie involved damming underground rivers and didn't mention aquifers.

The Guarani is also the locus and focus for numerous conspiracies and, more benignly, misconceptions. Among the latter is the assertion, in an Earth Institute paper, that because the Guarani covers an area the size of Texas and California combined, it "contains enough freshwater to sustain the world's population for two hundred years, and as water shortages affect us all in the future, the Guarani aquifer could be a lifeline for millions."[11] Left unstated, for obvious reasons, is the how-to of getting this water from Point A, Paraguay, to Point B, say, Yemen, where they need it most. This is a classic example of the error of considering the world's water woes as a single global crisis instead of as a series of regional ones.

The fact that the World Bank put some money into studying the Guarani is sometimes advanced as evidence enough that rich-world countries were attempting to assert control over the resource. As a result, the many NGOs and academics who were suspicious of foreign involvement in aquifer management lobbied Mercosur, the regional trade block analogous to the EU, to fight "any foreign control over the Guarani waters." Ridiculously, they were lobbying against World Bank involvement, though in this case, the bank was trying to bring about exactly what the NGOs said they themselves wanted: cooperative local control. In any case, a World Bank grant helped the four Guarani states to study the issue, and in 2010 the four signed the Agreement on the Guarani aquifer, the first shared-management aquifer concord in Latin America.

The agreement is interesting in several ways. For one, it was the first such signed after the UN adopted its Law of Transboundary Waters, and the Guarani text follows the UN's suggestions closely. Also, it is remarkable because even in an area where states are prickly about boundaries, there have been no real conflicts over the Guarani groundwater resource. Cynics might say this is because they really hadn't assessed the resource very well to this point. But it is interesting also for the wide range of participants — the governments concerned, academic research networks, the World Bank and the Organization of American States, some NGOs, and private companies, many of which were hoping to get a part of whatever action resulted. Not everything went smoothly: a proposed commission to assess the various national water resources never happened, and a proposed further agreement for joint management of the Guarani was quietly forgotten.

A study by two São Paulo academics, Pilar Carolina Villar and Wagner Costa Ribeiro, pointed to at least some of the agreement's strong points: "Equitable and reasonable use of water" was included in Article 4, in which states "shall promote the conservation and environmental protection of the Guarani Aquifer System so as to ensure multiple, reasonable, sustainable, and equitable use of its water resources." And, second, the obligation not to cause harm was included in Article 6: "Parties that perform activities or work for utilizing the water resources of the Guarani Aquifer System, in their respective territories, shall adopt all the necessary measures to avoid causing significant harm to the other Parties or the environment."[12]

It remains to be seen how these work in practice. But so far so good.

All in all, the global groundwater picture is dire but not desperate. In places it is, but even there, remedial actions can be taken, pushed either by necessity, foresight, better planning, or thrifty use. Even in California, best guesses are that the state could reduce demand by 17.3 cubic kilometres a year by proper management and conservation initiatives — enough to make up for the shortfall caused by the current drought.

Whether this *will* be done — with similar measures taken in China, the Middle East, and the rest of Asia — remains to be seen, but it is possible.

3
Rivers and Lakes in Trouble

The poet-philosophers who compiled the book of Ecclesiastes had it right, at least for their time:

One generation passeth away,
and another generation cometh:
but the earth abideth always...
All rivers runneth to the sea,
yet the sea is not full...

Alas, it is no longer true. The earth *abideth* okay so far, but not all rivers *runneth* any longer to the sea, and of those that do, a depressing number *runneth* a mighty load of crap into the sea. And unless we are very careful, the sea will indeed become full, and our species will be obliged to *maketh our way* to high ground to flee the rising waters that we ourselves have caused.

Water isn't as simple as it used to be.

It's easy to think there is an immense amount of water in the rivers and lakes of our planet. We look at any one of the Great Lakes and we can't see the other shore. Lake Baikal is more than a kilometre and a half deep. The outflow from the Congo is so strong and so steady that you can dip a bucket into the ocean a hundred kilometres off shore and drink fresh water (the "plume" has been known to stretch eight hundred kilometres out to sea). We read that the Amazon basin accounts for a quarter of planetary runoff all by itself. Many of the numbers are familiar: Russia's Baikal alone accounts for one-quarter of all the world's lake-held fresh water (twenty-three thousand cubic kilometres). Africa's Lake Tanganyika is second in volume (nineteen thousand cubic kilometres), and Lake Superior, on the US-Canadian border, is third at twelve thousand cubic kilometres. The North American Great Lakes, the world's largest lake system, account for 27 percent of global lake volumes.

We "know" there's a lot of water in these lakes and rivers because, unlike the aquifers that provide our groundwater, we can *see* them. We drink them, irrigate with them, play on and in them, drown in them in flood times, fret about them in dry ones, and use them, too often, as natural sewers. They act as drainage channels, provide habitat and nourishment to wildlife, provide transportation routes, and produce electrical energy. But the freshwater lakes and rivers from which we draw so much nourishment contain only about ninety thousand cubic kilometres of water in aggregate, a trivial 0.36 percent of the world's total supply of fresh water that is itself less than 3 percent of global water supplies. The rest, as we saw in the previous chapter, is water underground.

Rivers and lakes are easily used up, easily polluted. As we are now seeing a little more clearly, to our sorrow.

No exhaustive catalogue of rivers exists. It would be a pointless exercise: What is a river? A creek, a brook, a stream? Is a swamp or a bog a slow-moving river? I was once walking along a dry "riverbed" with a film crew in Namibia and we had to skip out of the way as a wall of water rounded a bend half a kilometre ahead of us — it had rained somewhere a long way away "upstream," though the skies were cloudless where we were filming. Well, the "wall" was only a foot or so deep, but suddenly there was a river where none had been a few minutes before. It was a standing joke where I grew up, in the arid interior of South Africa, that for eleven months of the year you could jump into a river and have to dust yourself off. Is a flood a temporary river?

Many geography texts insist that there are 165 rivers in the world classified as "major," but definitions are elusive and the number remains disputed. There are at least a quarter of a million rivers in Canada, many more if you count the unnamed links between adjacent lakes. The United States has nearly as many, Russia twice as many. Some facts are commonplace and in every grade-school atlas: the Nile is the longest river in the world (a shade longer than the Mississippi, which is second); the Amazon has the greatest annual flow (the Ganges is next, followed by the Congo). Of the twenty-five largest rivers of the world, three are in Africa, four are in South America, eleven are in Asia (if you count Siberia), five are in North America, and two are in Europe.[1]

Lakes, similarly, remain uncatalogued and oddly underappreciated in the water world. For example, the World Water Vision, issued by a slew of government and non-government organizations in the year 2000, mentioned lakes not at all, not even in the index, though lakes make up almost 90 percent of the world's non-frozen surface fresh water.

A credible estimate is that there are 117 million lakes greater in area than two thousand square metres.[2] They take up almost 4 percent of the earth's land area not covered by ice and are overwhelmingly concentrated in northern regions, where glaciation gouged countless holes in the earth's crust.

The twenty-eight largest account for 85 percent of the volume of all lakes worldwide. I have already mentioned Baikal, Tanganyika, and Superior; rounding out the top ten are Malawi-Nyasa, Vostok (in Antarctica), Michigan, Huron, Victoria, Great Bear Lake (in Canada's Northwest Territories) and Issyk-Kul (in Kyrgyzstan). The largest lakes in Europe are in Russia (Ladoga and Onega), and in South America (Maracaibo, Titicaca, Poopó, and Buenos Aires). All the Middle Eastern lakes of consequence — the Van, the Tuz, and the Beyşehir — are in Turkey. The Dead Sea, the largest lake in the Levant, is more saline than the oceans; Lake Kinneret (the Sea of Galilee) is not far behind in volume. Australia has no lakes to speak of.

Rivers and lakes intersect in complex ways. One small example is the Mekong River basin, in Southeast Asia. From mid-May to mid-October, the rainy season, the flow of the Mekong itself becomes so large that the delta can't support the volume. The water then backs up the Tonle Sap River to fill Tonle Sap Lake and its associated flood plain. This infusion has historically created one of the most productive fisheries in the world — or did so until recently. The Mekong, like so many other rivers, is in trouble.

Many rivers and lakes are in trouble either from overuse or from toxins that we have dumped into them. China's Yellow River is perhaps the starkest cautionary tale. The Yellow is "the cradle of Chinese civilization," in the comforting cliché of the dreamier propagandists, and what has happened to the Yellow can be taken as a metaphor for what is happening to rivers globally.

The Yellow is also known as the Huang He, the Hwang Ho, the Mother River of China, and sometimes just as "the river," its central reaches nourishing the plains that were, indeed, the locus of the birth of Chinese civilization. It is the second longest river in China after the Yangtze; it rises in the Bayankala mountains in Qinghai Province in the far west, flows through nine provinces, and empties into the Bohai Sea, 5,463 kilometres from its source — its basin covers almost three-quarters of a million square kilometres. The water of the river is indeed yellow because it contains a vast amount of silt (if you scoop up a bucketful, as much as 60 percent of the contents by weight is fine-grained yellow silt). The silt is picked up by the rapidly flowing waters pouring through deep canyons carved into North China's great Loess Plateau.

The Yellow has other names too, among them China's Sorrow, for the Huang He floods often, and it floods massively. The loess silt in the river, 1.6 billion tonnes of new sticky yellow mud every year, is one reason it has become so deadly. Over the years it has deposited thick layers of silt along the riverbed as it flows through central China's plains, with the curious effect of raising the riverbed itself, often to levels higher than the surrounding flat plains. Thousands of kilometres of dikes have been built over the centuries to contain the river, but in sharp floods they often give way (or are overtopped), with catastrophic results. The flatness of the plains means that every flood covers hundreds or thousands of square kilometres of heavily populated land. The history books have counted 1,593 episodes of flooding in the last four thousand years. The worst flood of all killed four million people in one miserable episode, still one of the greatest natural disasters in human history. That was less than a hundred years ago.

Flooding is not so much a problem now. Not much more than a quarter century ago, as Lester Brown, then head of the Worldwatch Institute, wrote:

> With more and more of its water being pumped out for the country's multiplying needs, the Yellow River began to falter. In 1972, the water level fell so low that for the first time in China's long history it dried up before reaching the sea. It failed on 15 days that year, and intermittently over the next decade or so. Since 1985, it has run dry each year, with the dry period becoming progressively longer. In 1996, it was dry for 133 days. In 1997, a year exacerbated by drought, it failed to reach the sea for 226 days. For long stretches, it did not even reach Shandong Province, the last province it flows through en route to the sea. Shandong, the source of one-fifth of China's corn and one-seventh of its wheat, depends on the Yellow River for half of its irrigation water.[3]

The water that is left is in parlous condition, much of it unfit even for irrigation, much less for drinking. The Yellow estuary is a cesspool, and nearby offshore waters are not much better.

Although it is perhaps the most visible manifestation of water troubles in China, the drying-up of the Yellow River is only one of many such signs. Satellite photographs show hundreds of lakes disappearing and local streams going dry in recent years as aquifers diminish and springs cease to flow. As water tables have fallen, millions of Chinese farmers are finding their wells pumped dry. In the 1950s, the country had fifty thousand rivers. True enough, many of them were small, with catchment areas not much more than one hundred square kilometres. Even so, by 2010, the number was down to twenty-three thousand — the Chinese had "lost" twenty-seven thousand rivers.

Even those not lost were suffering. Four-fifths of the remaining waterways are so polluted they no longer support any fish — or at least not any fish you would want to eat. Even in the Yangtze, the only river in China greater than the Yellow, the fish catch has declined by more than half just since the 1960s. The government itself has indicated that fully 70 percent of all waterways had been fouled to the point of being unsafe.

By China's own estimate, 280 million Chinese are obliged to drink unsafe water. Fully a quarter of municipal water-treatment plants in the country do not come anywhere close to complying with quality-control standards. Even more dismally, every year some 190 million people in China fall ill (and 60,000 die) from diseases, such as liver and gastric cancers, caused by water pollution.[4]

If only China were unique, but it is not. Everywhere you look, rivers are in similar trouble. Four out of five humans now live in areas where river waters are highly threatened by pollution, mostly in Asia but also in Africa and Latin America. Throughout Asia, rivers routinely carry three times more fecal coliform bacteria than is deemed safe, and demand for clean water far outstrips supply. As a piece in the *Journal of Health, Population and Nutrition* put it in 2008, "So common is water contaminated with human feces throughout South Asia that it is accepted as the norm...available drinking water [contains] organisms whose ecological niche is the human intestine.... Those who can afford it buy bottled water (of dubious quality), and the majority are left to drink the available contaminated water."[5] The Mekong River that supplies drinking water, farming irrigation, industrial water, the fishery, and wastewater disposal for half a dozen countries has seen its quality deteriorating for decades. A 2003 report by Wijarn Simachaya for the Mekong River Commission Secretariat declared that "economic development of the basin has resulted in elevated levels of pollution from both point and non-point sources. Degradation of water quality in parts of the basin has evolved gradually over time until eventually becoming apparent and measurable." Simachaya's charts showed dangerous levels of parasitic organisms and bacteria, chemical and industrial pollutants, and farm waste.[6]

In Latin America as late as 2010, only a few percentage points of human waste was being treated in any way instead being dumped on the land (and therefore into the water tables) or into waterways — as we saw with São Paulo. Even the hitherto pristine waters of Lake Titicaca are being threatened by sewage and industrial waste, much of it from the poor adjunct to La Paz called El Alto. The waters off Rio de Janeiro became briefly notorious in 2014 when sailors practising for the 2016 Olympics were advised not to fall overboard lest they become ill from the water or bump into dead dogs. And the Tietê River than runs through São Paulo and Recife actually got worse from 2000 to 2010, despite a $400 million capital infusion from the World Bank and the federal savings bank.

In Africa, all the major rivers have seen cataclysmic drops in fishing catches. The rapid growth of urban populations has far outstripped the capacity to deal with human waste, polluting what meagre water sources there were. This leads to using polluted water for irrigation, and to widespread outbreaks of cholera, dysentery, and other water-vector diseases. About half of Africa's countries cannot supply clean water to at least half of their populations.

In terms of the number of people it kills, "dirty" water is the world's most serious pollution problem.

North America and Europe

Mark Mattson, an environmental lawyer and Riverkeeper, pointed out to a small audience in Vancouver in 2014 that "we don't clean up the rivers or beaches anymore — we just post No Swimming notices, as though that were *normal*, and think nothing of it." True enough, every summer there are Do Not Drink and Do Not Swim notices in thousands of waterways across North America — and there is virtually no indignation. We don't demand that the polluters clean up what they have wrought. We don't seem to demand anything very much.

A small example of how this works in practice is a gold-mine tailings pond breach in a pristine corner of British Columbia in August 2014. Hundreds of thousands of cubic metres of "water" containing high concentrations of arsenic, lead, and mercury decanted into the Cariboo District waterways, soon making its way down to the Fraser River, a major arterial river known for its salmon runs. The president of the mining company, Brian Kynoch, was distraught. "A gut-wrenching experience," he called it. But he was also mightily surprised. "If you asked me two weeks ago if that could happen, I would have said it couldn't happen," he said at a media update in a community hall in Likely, BC. "I know that for our company, it's going to take a long time to earn the community's trust back."[7] The provincial government's response, at least at first, was eerily similar: "We'll make sure it does not happen again." This was similar to the response that greeted the leak from a ruptured storage tank outside Charleston, West Virginia, that contaminated the water for some three hundred thousand residents. The cause there: a chemical used in the processing of coal. The result: Do Not Drink and Do Not Bathe warnings. The consequence? Oh, sorry, we won't do it again.

True, a few months after the Mount Polley Mine spill, the BC government astonished activists by inviting public opinions on what had caused the spill: "There are a lot of opinions flying around in the public domain, perhaps even in the private domain, and we thought to be properly diligent that we should invite anyone who cares to, to formalize those views and get us to think about them," said Norbert Morgenstern, an engineering professor emeritus at the University of Alberta and chairman of an investigating panel. Even so, the solicited opinions were restricted to the "mechanism" of dam failure at the tailings storage facility, with a view to heading off future such failures.[8] No one was asked to comment on the *why* of the spill, or on whether the tailings pond should have been there in the first place.

If you're strolling along a stream bank now almost anywhere in the developed world, and most especially near habitation, would you dip your hand into the water and take a drink? Of course not. You would have no idea what the water contained, or who put it there. It could be anything, or anyone. The presumption that the water is clean, and that the people who use it would keep it that way, long ago began to seem quaintly old-fashioned. Even in Nova Scotia, the bucolic small province where I now live, a small place that is underpopulated and under-industrialized, covered mostly by trees — even here, people no longer drink water from rivers and lakes. "Surface water should never be used for drinking," the director of environmental health with the provincial Health and Wellness Department told the media in August 2014. Gary O'Toole was commenting on a high fecal coliform count in local rivers. "Those results are not unexpected," he said. "Our standard advice is that people should never drink water taken directly from a lake or river."[9] Of course, those results *might* be because of excessive bear shit in the water. But it is much more likely that family farms are the culprit. Not massive agribusinesses, either — just little farmers, all doing their best.

The people who drink bottled water are generally looked at askance by environmentalists, and for good reason: the millions of tonnes of plastic those bottles are made of are a ghastly presence in the environment. But the critics have forgotten at least one of the reasons people carry bottled water: they don't trust the water supplied to them.

Poisonous algae are found in rivers and lakes all across North America, fed by nitrate and phosphorous runoff from farm fertilizers, porous municipal sewage systems, and home septic "tanks" often made from rusting old oil barrels. Just this sort of accumulated runoff down the Mississippi River has caused a dead zone in the Gulf of Mexico of more than twenty-thousand square kilometres. Almost every other estuary on the continent — Chesapeake Bay, Sacramento Bay, the Columbia River estuary, the St. Lawrence, and many more — suffers from the same effect. Poignantly, the Amish farmers of Pennsylvania, famed for their resolute rejection of modernity in all its forms, have been fingered as a major villain in the pollution of Chesapeake Bay, runoff from uncontained manure piles the given reason.

Decades after the Clean Water Act, the American water situation looks like this:

- Rivers and streams: of the 28 percent, or 1,632,721 kilometres the Environmental Protection Agency (EPA) assessed, 837,876 kilometres were found impaired, and 11,756 threatened. That is slightly more than half that were still polluted.
- Lakes and reservoirs: with 43 percent assessed, 4,885,001 hectares were found impaired, of a total 7,265,227, or 67 percent still polluted.
- Bays and estuaries: of the 37 percent assessed, 61,157 square kilometres out of the 85,449 were polluted, or 71 percent.
- Coastal shoreline: of the small 1.4 percent assessed, 11,734 kilometres out of 13,591 were polluted, or 86 percent.
- Ocean and near ocean water: only 3.1 percent was assessed, and 2,743 out of 4,343 square kilometres were polluted, or 63 percent.
- Wetlands: only 1 percent was assessed, and of those 450,187 hectares, 217,920 were polluted, or 48 percent.
- Great Lakes shorelines: a good percentage was assessed (85.2 percent), and 7,005 kilometres out of the 7,131 were polluted, or 98 percent — that is, the EPA found only 126 kilometres of the American Great Lakes shoreline "good."
- Great Lakes open water: 88.1 percent was assessed, and 137,969 square kilometres out of 138,129 were polluted, or as close to 100 percent as makes no difference.

This actually counts as progress — almost half the country's rivers and streams are in reasonably good health, and about a third of its lakes and reservoirs. True, the coasts didn't fare so well (Georgia to Massachusetts the worst, then San Francisco Bay and the Pacific from Los Angeles to Mexico), and the state of the Great Lakes is still abysmal, though it is improving. The worst polluted places are still Lake Erie, southern Lake Huron, Georgian Bay, Lake Ontario, and the lower St. Lawrence estuary.

In a way, though, the numbers may be better than they look. Chemical and industrial pollution is much lower than before. Much of the current pollution stems not from factories, at least not so much in the United States, but from the actions of ordinary consumers, and from farming. As I've pointed out elsewhere, there are now so many people that ordinary human actions are causing major pollution problems (the best-known "ordinary human action" is a curiosity: suntan oil from millions of bathers is now a serious pollutant on Mediterranean beaches).

Canada, with its small population and vast water resources, should have better water health than the United States, and it does, but not to the degree expected. The Commission for Environmental Cooperation, a trilateral monitoring group, has rated only 44 percent of southern Canada's fresh water as excellent or good. The standard was considered as fair at 33 percent of sites monitored, and marginal or poor at 23 percent of them. In Canada, phosphorous remains an issue — phosphorous levels exceed guidelines at more than half the sites monitored.

Twenty years ago, it was a national disgrace that two major Canadian cities, provincial capitals on both coasts, had no sewage treatment facilities at all but simply dumped the raw effluent into the oceans. Since then, Halifax in Nova Scotia has built an expensive treatment plant, which mostly works. Victoria, British Columbia's capital, however, still pumps somewhere between 82 million litres and 130 million litres of sewage daily into the Juan de Fuca Strait. The stuff is pushed along two underwater pipes by twin thousand-horsepower pumps, and emerges into the ocean a kilometre offshore, sixty metres below the surface.

Apologists say the ocean acts as a natural toilet that disperses waste with minimal environmental impact. An indignant website called Responsible Sewage Treatment Victoria, whose purpose is to argue against building an expensive treatment facility, points out that the city's sewage is screened before it emerges into the ocean, "so there are no floaties," as though that makes everything hunky-dory. In the 2012 federal by-election in the Victoria riding, all the candidates except the winning New Democratic Party member opposed building a treatment plant, even the Greens (they were for treating the nasty stuff, just not in the way the project's proponents were suggesting). The only person for it, it seems, is the governor of Washington, where some of those non-floaties end up, who sent an irritable letter early in 2014 to the province's premier, pointing out that "it is now more than 20 years since your province agreed to implement wastewater treatment in greater Victoria, and yet today Victoria still lacks any treatment beyond screening. Delaying this work until 2020 is not acceptable."[10]

In Canada, in some ways, things are getting worse. For example, for Mark Mattson, the environmental lawyer, a raft of environmental protection laws enacted in the last decade of the twentieth century allowed him to become the first litigator in Canada to successfully prosecute a polluter, only to have every one of those laws gutted by the first prime minister of the twenty-first, Stephen Harper.

Mexico's water can sometimes be clean, but not very often. Even the ecotourism resorts on the Pacific coast can do little to avoid the toxins pouring down on them from rivers flowing from the interior. Mexico City's water can emerge from residential taps in a variety of sometimes strangely attractive colours, but its effects on gastrointestinal tracts is not so pretty. As the *Guardian*'s Kurt Hollander once put it, "Although the excrement that I and millions of others dump each day into toilets throughout Mexico City takes an amazing voyage beneath the city streets, through 6,000 miles [9,655 kilometres] of pipes, 68 pump stations and across almost 100 miles [160 kilometres] of canals, tunnels, dikes and artificial lakes, it has an uncanny knack of finding its way back to me."[11] The Commission for Environmental Cooperation has found fecal coliform bacteria, fed by human or animal waste, in more than half of the drinking water supplied to Mexicans. And, "as in the rest of North America, levels of nitrogen and phosphorus in surface water are also a problem for Mexico. Elevated levels of pollutants containing these elements were detected at a majority of monitored sites."

Europe has historically not been much better. Since the Industrial Revolution, Europe has treated its waterways as convenient conduits of waste to the sea, destroying in the process the biodiversity of thousands of kilometres of rivers and polluting coastal waters, never mind what it has done to the humans obliged to drink the water along the way. In the days of the Industrial Revolution, just dumping stuff into waterways seemed sensible enough — Europe has several million kilometres of flowing water and more than a million lakes and, like the oceans, they seemed inexhaustible. As a result, stinking rivers, dying fish, and polluted lakes were commonplace.

But in the last twenty years or so, sewage and industrial wastes have been sharply reduced across the continent, resulting in what the EU calls a "measureable improvement in water quality." Phosphorous levels and organic wastes have been reduced at source, and the amount of discharge into the oceans has dropped. Heavy metals and other toxins are still present, but at levels now harder to detect.

European agriculture, however, remains unreconstructed and to a degree recalcitrant, an arrogance due at least in part to the high esteem in which it has been generally held in Europe, and to the high levels of subsidy to which it has been accustomed. In essence, farming has been a coddled darling, and it has behaved uncaringly because of it. The levels of nitrates from farming runoff have actually increased in the last decade, as an EU report asserted in 2013: "Nitrate pollution, particularly from fertilizers used in agriculture, has remained constant and high. Nitrate concentrations in rivers remain highest in those western European countries where agriculture is most intensive."[12] At the turn of the twentieth century, more than eight hundred pesticides of varying virulence were licensed for use in the EU. That number has been dropping, and the tonnage spread has diminished too, but mostly because newer pesticides are more aggressive than older ones. The picturesque but unhelpful practice of "blowing" liquid manure onto winter snow cover is still prevalent, leaching bacteria into the waterways.

Despite efforts to clean up industrial sites, twelve European countries have reported heavy-metal contamination of groundwater because of mining dumps and industrial discharges.

In some ways, you'd expect lakes to be in better shape than rivers. Unlike rivers, they don't just pass through on their way somewhere else, so if the water levels change, or the pollution quotient increases, it would be obvious to the users, and they would be impelled to do something about it.

Unless they felt powerless to do so. Or didn't feel like it.

Most of the world's major lakes are in an okay condition, but there are worrying signs. Baikal, the deepest lake on the planet at 1.7 kilometres, is beginning to suffer surface pollution from surrounding pulp mills, as well as "cultural eutrophication," the technical term for human abuse. (Still the native seal population has been making a comeback, thanks largely to Living Lakes, an NGO operated out of Lake Constance in Germany.) But even a lake this large is not immune to input pressures. Lake Hovsgol in Mongolia is the headwater for Baikal, and its supply (and quality) is now under substantial pressure from resorts and hotels, among other things. Lake Tanganyika is showing even more distressing signs of eutrophication — there are too many people dumping too much sewage sludge into water that seemed limitless but wasn't. Worldwide, the trend is the same. Almost twenty years ago, UNEP, the UN Environment Programme, did a survey on eutrophication and found that 54 percent of lakes in Asia were eutrophic, 53 percent in Europe, 48 percent in North America, 41 percent in South America, and 28 percent in Africa. There is little reason to suggest that things have improved since then.[13]

A long-running study by researchers at the Experimental Lakes Area in Ontario has found something depressing: the notion that eutrophication can be reduced by controlling the amount of nitrogen entering lakes, a policy on which the EU and others are spending many millions of dollars, may actually be making things worse rather than better. The study was done on a small lake, Lake 227, in the Precambrian

Shield area of Ontario. The scientists, led by David Schindler of the University of Alberta, fertilized the lake for thirty-seven years with constant annual inputs of phosphorous and decreasing inputs of nitrogen to test the theory that controlling nitrogen was sufficient to control eutrophication. For the final sixteen years, the lake was fertilized with phosphorous alone. The lake, Schindler reported, remained stubbornly highly eutrophic: "The impact on human society is immense, as cultural eutrophication severely reduces water quality, which not only kills and contaminates fish, shellfish and other animals, but also can become a health-related problem in humans once it begins to interfere with drinking water treatment."[14]

In an all-too-common act of scientific vandalism, the increasingly anti-science national government of Stephen Harper shut the Experimental Lakes Area facility down, with no warning and no attempt to salvage its data, possibly because Schindler has long been an opponent of some of Harper's cherished projects, including the Alberta tar sands. The facility was rescued only through a last-minute grant from the Ontario government.

The American Great Lakes, as mentioned before, are far from pristine. But they are improving in many places. However, in the summer of 2014, the residents of the Ohio city of Toledo were told not to drink their water — "sludgy algae" the given reason, its cause said to be "unknown." Toledo is on the shores of Lake Erie, the smallest, shallowest (average depth only eighteen metres), and most intensively used of the Great Lakes. It has been declared dead once before, in the 1960s, and was revived through the US's Clean Water Act. More recently, Lake Erie has been dying again because of increased population pressures and the over-application of much more potent phosphorous-based fertilizers. The mechanism is not at all unknown, as the residents of Toledo had been told. On the contrary, it is well understood: the phosphorous doesn't just fertilize farm fields; it also "feeds a poisonous algae whose toxin, called microcystin, causes diarrhea, vomiting and liver-function problems, and readily kills dogs and other small animals that drink contaminated water. Toledo was unlucky: A small bloom of toxic algae happened to form directly over the city's water-intake pipe in Lake Erie, miles offshore."[15]

All the other Great Lakes are cleaner than they were thirty years ago, when beluga whales tagged in the upper St. Lawrence River met the classic definition of toxic waste. This is largely because of the International Joint Commission (IJC), the bilateral body tasked by Canada and the United States with managing the lakes' ecosystem. Still, in 2002, fish caught on the Great Lakes contained many neurotoxins, including PCBs and methylmercury, and many studies have indicated that persons consuming such fish were vulnerable to a shopping-list of negative side effects. The problem is that scientists have only a hazy notion of what chemicals these fish actually contain. Researchers for the IJC have only been able to identify about a third of them: "Several halogenated compounds as well as antibiotic and other pharmaceutical residues in Great Lakes samples remain unidentified. The presence of brominated diphenyl ethers, chlorinated paraffin and napthalenes and PCB metabolites in the tissue of a variety of species ranging from snapping turtles and herring gulls to polar bears and humans, remains a mystery." Worse, because they can't identify the chemicals themselves, there is no way to guess at what chemical mixtures might do to animal tissue. The chemical soup ingested by fish, and thus by humans, has unknown properties: "Some of these chemicals interact with each other, but how mixtures affect the biota remains unknown."

In 2010, an IJC report said that PCB concentrations in Great Lakes water were substantially down, but were still a hundred times higher than they should be.[16]

Quality of the water is not the only issue: the sheer volume of water, or the lack of it, is another. The Great Lakes may house a quarter of the world's lake-held fresh water, but, as pointed out earlier, they are a long way from delivering even a fraction of the world's renewable water. Even a small drop in inflow causes the water levels to drop. In fact, in June 2013, mean levels dropped to the lowest levels recorded since measurements began in 1918 — annoying cottagers and recreational boaters, but much more alarming for the $3- billion-a-year Great Lakes–St. Lawrence Seaway shipping industry. Prolonged drought was blamed.

By November, just a few months later, the Army Corps of Engineers was reporting a different story: water levels had recovered substantially, because of greater than normal winter snows and heavy spring rains. They weren't back to "normal" (however that is measured), but they did rise fifty centimetres or so.

There is also, happily, good news about rivers.

If all you know of the Hudson River is the stretch that skirts New York City, you will know it mostly as an unlovely thing, rank-smelling, ripe with flotsam, and in heavy rainfalls the recipient of millions of cubic metres of stormwater mixed with "partly treated" sewage. To get the flavour of that particular stretch of river, you could do worse than read a queasy-making piece by Lindsay Crouse of the *New York Times*, about triathletes paddling and then swimming down the river, collecting along the way what the athletes call the Hudson Mustache, "the thick band of silty debris that clings to a swimmer's upper lip." Crouse quotes one of the participants, who likened the first wave of swimmers to emerge from the river to coal miners, their faces obscured by grime, the only "clean" skin where the goggles had been.[17]

Well, okay, that doesn't sound like good news, and it isn't. But upriver, things are different.

The Hudson was long considered the region's sewer, but in the 1960s the fishermen along the river began to fight back. At the same time, a group of citizen activists launched an effort to save Storm King Mountain from the clutches of the electrical utility, Consolidated Edison, that wanted to build a hydro plant across one of the most scenic stretches of the river. This resulted in the first true test of the statutes recently passed by US Congress, the most important of which was the National Environmental Policy Act. The activists, under the umbrella of the Scenic Hudson Preservation Coalition, launched a lawsuit against the utility in 1962. It dragged on until 1979, when Con Ed finally capitulated.

Here's how the Riverkeeper website describes what has happened since:

> Three decades later, the Hudson has once again regained its status as the region's gem. Anglers, boaters and bathers flock to its waters to experience the wonders of this great river. By and large, industries and municipalities have ceased their polluting ways and have developed a respect for the resource. However, sustained vigilance is needed to ensure that the great gains in water quality are not reversed. Many river segments, particularly in urban communities and areas of sprawl growth, remain threatened. And with government enforcement of environmental laws spiralling downward at the state and federal level, Riverkeeper is working overtime to bring violators to justice.

While infinitely better, the cleanup is not yet finished. GE still has a nasty reservoir of PCBs to clean up; there is still what the River-keepers call the "toxic brew of sewage, coal tar, PCBs and heavy metals in the Gowanus Canal."

It is true that the Hudson activists benefited from one resource few other such groups can claim: true star power in the person of Robert F. Kennedy Jr., an environmental lawyer who has been Riverkeeper's driving force for more than two decades. Kennedy can sometimes hold opinions that border on the crank (he's suspicious of vaccinations, for example), but the Kennedy name, and money, still effortlessly attract attention.

Elsewhere in the United States, the Love Canal has been cleaned up. The Clean Water Act and the Clean Air Act in the United States have provided regulators with some teeth. A survey by the US Geological Survey in the summer of 2014 showed that restrictions on pesticide use have already had an effect — pesticide residues in American streams and rivers have dropped sharply.[18]

In the EU, similarly, there has been progress. The EU Water Framework Directive, adopted in 2000, sets out several broad principles governing water pollution:

- The level of protection should be "high." In this context, this means that countries are directed not to settle for minimum acceptable levels (those at which human health is *probably* not affected). "Water" here means all water resources and natural ecosystems.
- Actions should be governed by the precautionary principle. That is, policy should be based on recognized scientific knowledge but should leave a margin for error, and err on the side of caution where the basic science is not fully established or knowledge is incomplete.
- All states have a moral duty to prevent damage to the environment. This is called the prevention principle, which recognizes that it is more difficult and more expensive to treat pollution after it has been committed than it would be to prevent it at source.
- The polluter pays principle must be enforced. This principle is simple enough: those who produce wastewater or contaminate the environment are obliged to pay the full cost of remediation. The EU argues that invoking this principle prevents distortions in the marketplace by ensuring that external costs are included in production costs. It also acts as an incentive to prevent pollution at the source.
- Rectification should be at the source: wherever possible, pollution should be fixed where it is committed, rather than downstream.

The most famous river rectification effort to date has been with the Rhine, Europe's most important waterway. Not very long ago, the Rhine was derided as "the sewer of Europe," but it has been steadily cleaning itself up for several decades. There are now fish in the river, though you wouldn't necessarily want to eat them, and in places it is even possible to swim without hazard. The Rhine is the poster child for awakening awareness of how rivers went wrong — but also for how they can be fixed.

The Rhine rises in the Reichenau municipality above Lake Constance in Switzerland and flows for 1,320 kilometres to the Wadden Sea in the Netherlands. Most of its course is through Germany, but its catchment area of 185,000 square kilometres takes in large swathes of Germany, France, Holland, Austria, Luxembourg, Liechtenstein, Belgium, and even Italy, and contains more than seventy million people.

By the 1980s, the Rhine was not really a river anymore, but an engineered shipping lane and a conduit for a multiplicity of poisons. Dutch law professor Hans Ulrich Jessurun d'Oliveira, usually known just as Ulli, has been involved in the cleanup effort since the late 1970s. "Everything was wrong with the river," he has said. "It was considered dead. Nothing could live in it. Enormous industries were all based on its banks because of the possibilities it offered for getting rid of waste." The biggest villains were a group of French potassium mines that poured hundreds of tonnes of waste salts into the river every day, killing the few fish that remained. A group of Dutch tulip growers, who found they could no longer use the Rhine's water even for irrigation, formed the Clean Rhine Foundation and sued. Ulli was the group's chairman. As he explained, "It was a French government-owned company, Les Mines de Potasse d'Alsace, and they dumped chlorides in the Rhine in such bulk — that was phenomenal. It was the biggest polluter — in terms of bulk — in the Rhine. And so we thought, this is a target we could use in order to open the eyes of the people [to see] that this is really a threat to Dutch society."[19] In 1988, after the case reached the Netherlands Supreme Court and the European Court of Justice in Luxembourg, the mines were ordered to pay compensation — unspecified. But the dumping continued unabated.

It took a catastrophe to get the riparian countries to pay real attention. What followed is well known in European environmentalist lore.

In October 1986, a fire broke out in an electrical switching box in a riverside warehouse in Basel, Switzerland, nearly five hundred kilometres upstream from the place where the Sieg enters the Rhine. It was no one's fault — a mechanical system had failed. But flammable material was stored nearby, and that was definitely someone's fault — the owner's, Sandoz, one of Switzerland's largest chemical companies. Through a cascade of sloppiness and bad luck, more than thirty tonnes of poisons poured into the river following the fire, an evil brew of herbicides, fungicides, pesticides, dyes, heavy metals, and two tonnes of mercury. Yet even this ecological disaster, perversely, was leveraged into good news, perhaps because Chernobyl was still a recent and terrifying memory. The people and their politicians were frightened into action.

The Rhine cleanup, begun with little popular enthusiasm decades before, was galvanized. Riparian countries would, at last, do what must be done. As little as four years later, dozens of dams and other obstacles had been removed, wetlands had been restored wherever possible, and thousands of salmon fingerlings had been released. By the turn of the century, several viable populations of salmon were spawning in the Rhine, a fraction of what was there before. Not yet edible, but self-sustaining nonetheless.

"The rebirth of the Rhine ... has to count as one of the great environmental success stories of the century," said Angela Merkel, then Germany's environment minister, with considerable hyperbole but some justification.

Pollution can, it is clear, be fixed. All it takes is political will. And money. And a determination to make the polluter pay.

Other rivers in Europe have shown similar improvements — Austria's Upper Drau, the Órbigo of Spain, and the Danube (though the Danube actually went from terrible to pretty bad and is still only on its way to okay).

Amiable cooperation along the Danube dates back to the 1856 Treaty of Paris, which set up a body called the European Commission of the Danube, with representatives from all riparian countries. It was concerned neither with monitoring supply nor with pollution — modern problems about which the commissioners were entirely innocent. Its purpose was to make sure that navigation along the river was free to all, and it ran smoothly enough (with hiccups for the world wars) until 1948, when the Cold War arrived and, as the website for the Strategic Action Plan now delicately puts it, "new alliances resulted in a new management approach," meaning that the East Bloc no longer wanted anything much to do with the West or any vessels emanating therefrom.

And there it rested until the mid-1980s, when alarms began to spread over the increasing degradation of the river. This had something to do with the work of a Russian researcher, Irina Zaretskaya, who was highly critical of the engineering works that had been perpetrated on the river, and even more critical of the quality of its water. Of the Danube's re-engineering, she wrote: "These works [carelessly] changed cross sections, coast lines, slope, bottom and suspended sediment discharge, as well as water quality... The quantitative and qualitative depletion of water resources in individual regions of the basin has resulted in a critical situation, especially during dry periods. An increasing water resource deficit in the region can become a brake on the economic development of the countries." As to the quality of the water, she found that dozens of cities and half a dozen countries allow huge amounts of "insufficiently purified storm runoff, industrial wastes, and agricultural pesticides to enter the river," virtually unmonitored. And then the zinger: The Danube poured 80 million tonnes of contaminated sediments into the Black Sea every average water-flow year.[20]

With this prodding, in 1985, the eight riparians (Germany, Austria, Slovakia, Hungary, Croatia, Serbia, Romania, and Bulgaria) signed the Declaration of the Danube Countries to Cooperate on Questions Concerning the Water Management of the Danube, fortunately for posterity renamed the Bucharest Declaration, that established a basin-wide monitoring network. This was followed by a meeting in Sofia, Bulgaria's capital, in 1991, that set up the Environmental Programme for the Danube River Basin, which in turn was succeeded by the

International Commission for the Protection of the Danube River. This remains the main instrument for cooperation among the Danube countries. The point of all these proliferating commissions was to make sure all the riparians were on the same page and used the same monitoring metrics, so that they could thereafter discuss the issue of liability for cross-border pollution while defining rules for the protection of wetland habitats and developing guidelines for conserving areas of ecological importance or aesthetic value.

Since then, water quality has improved, though it remains true that the river's water can only be used for drinking purposes on its highest reaches, between Dettingen and Leipheim in Germany and Mohács, in Hungary.

More interestingly, a private-public initiative called the Danube Water Program has been created, initially funded by the Austrian government and headquartered in Vienna, its declared aim being "to support the water supply and waste water sector in improving operational practice improvement within the Danube Region." This is a joint venture between the members of the International Association of Water Supply Companies, a private consortium that manages mostly wastewater issues in the basin, and the World Bank. The program coordinator is Philip Weller, a Canadian engineer and hydrologist. In recent surveys, a few pollutants, like organochlorine pesticides and heavy metals, have actually been getting worse over the last decade, but a great many others have been satisfactorily reduced, such as ammonium and phosphate levels and biodegradable organic pollutants. Efforts are, as they say, ongoing.[21]

We're polluting the oceans too, of course, as we have already discussed. It is true this has little to do with freshwater supply and management, except in the larger sense that the planet's health depends to a scary degree on healthy oceans.

A report in *Science* in January 2015 examined human impact on ocean life and concluded that "although defaunation has been less severe in the oceans than on land, our effects on marine animals are increasing in pace and impact. Humans have caused few complete extinctions in the sea, but we are responsible for many ecological, commercial, and local extinctions. Despite our late start, humans have already powerfully changed virtually all major marine ecosystems."

In an interview, the report's lead scientist amplified the point. "We may be sitting on a precipice of a major extinction event," Douglas J. McCauley, an ecologist at the University of California, Santa Barbara, told the *New York Times*'s Carl Zimmer. "But there is still time to avert catastrophe. Compared with the continents, the oceans are mostly intact, still wild enough to bounce back to ecological health."[22]

A 2006 UN Environment Programme report found that "overall, good progress has been made on reducing Persistent Organic Pollutants (POPs), with the exception of the Arctic. Oil discharges and spills to the seas have been reduced by 63 percent compared to the mid-1980s, and tanker accidents have gone down by 75 percent, partly as a result of the shift to double-hulled tankers."[23]

It is also good news that both rivers and lakes have acquired their protectors.

One way of tracking the progress being made in managing the world's rivers, for example, is through an Australian public-private initiative called the International RiverFoundation and its associated International River*symposium*. Launched by the Australian city of Brisbane and its energetic mayor, Jim Soorley, its declared aim is "to fund [through partnerships] and promote the sustainable restoration and management of river basins." That partnerships notion raised a few eyebrows in the prickly world of water NGOs, especially when Coca-Cola was enlisted as a partial funder of the foundation's inaugural European prize in 2013. However, the foundation remains unrepentant, pointing out that one of its declared aims is "to support the corporate social responsibility of companies by facilitating actions and projects in sustainable river basin management."[24]

Be that as it may, since 1999, the foundation has awarded annual RiverPrizes to acknowledge cleanup and remedial efforts in river basins worldwide. Winners have included England's Mersey Basin Campaign, Canada's Grand River Conservation Authority, Australia's Blackwood basin, Asia's multi-jurisdictional Mekong River Commission (a winner despite the troubles it is facing), the Siuslaw River system in the United States, the Drôme River in France, and others. The International Commission for the Protection of the Danube River won the 2007 prize, the Willamette River in Oregon won in 2012, and Kenya's Mara River Water Users Association won in 2013. The separate European RiverPrize, initiated in 2013, awarded the inaugural prize, predictably, to the Rhine.

Everyone agreed it was justified. It showed what can be done by clever management and political will.

There are dozens of NGOs and quasi-NGOs whose business is to monitor lake waters worldwide, to report on them, and to encourage remediation where necessary. In many cases, they partner with local governments. In just as many others, they operate without government participation or scrutiny.

One of the most interesting is Living Lakes International, originally a creature of the Global Nature Fund, a foundation based at Lake Constance in Germany. The network grew out of an informal alliance between four partners: the Lake Constance Foundation, the Mono Lake Committee in California, the Lake Biwa Environmental Research Institute in Japan, and the Wilderness Foundation, whose job is to take care of Lake St. Lucia in South Africa, an exquisite tropical jewel in northern KwaZulu-Natal Province, itself part of the iSimangaliso Wetland national park. These four were joined by the clumsily named International Lake Environment Committee Foundation (ILEC), a Japanese-based NGO that had produced a seminal report on how to manage lakes for sustainability.[25]

The initiative was subsequently adopted by the Chinese, who hosted an ILEC-sponsored World Lake Conference in Wuhan, responding to an alarmed call from the country's own growing environmental movement. The topic at hand: "Eighty percent of [Chinese] lakes are suffering from heavy pollution through industrial and domestic sewage, algae blooms . . . excessive water withdrawal, and increasing development. . . . The state of the world lakes has deteriorated alarmingly during the last decades, and Chinese lakes are a sad example of it. . . . The 24,900 Chinese lakes cover an area of over 80,000 square kilometres, and with a few exceptions nearly all lakes are heavily polluted or on the verge of drying up."

Chinese alarm about the country's lakes kicked into gear around 2000, when the government stopped pretending that the problems were only malicious propaganda from "wreckers and saboteurs," a nice phrase from early Leninist days. One of the first signs was the sudden attention paid to the attempt to save the failing Baiyangdian Lake system, the largest in North China, stretching across the flat plains that lie to the south of Beijing. In the 1950s, the lakes of Baiyangdian covered more than 800 square kilometres. By 2000, no more than 480 square kilometres remained under water, and a decade later, the number was down to 114 square kilometres.

As of 2014, the Living Lakes network had scooped up more than seventy partners representing fifty-three lakes worldwide, from all three North American countries, a handful of South American countries, a dozen or so in Europe, and others, from Kyrgyzstan to Estonia to Mongolia. The network is financed through fundraising efforts that include, encouragingly, grants from corporations like Daimler, Lufthansa, and Unilever. The representative lakes vary greatly in size, type, location, and health. They may be urban or remote. And they are chosen partly because they already have active local committees to monitor their health and partly through a more deliberate process of selecting lakes that contain, in the words of the Living Lakes people, "valuable ecosystems and hotspots of biodiversity, offering important ecosystem services such as drinking water, irrigation for agriculture, fish, recreation, buffer zones against floods and micro climates favorable for all kinds of cultivation as well as for the people living in the watershed."

The UN has helped the cause. Two lakes, Baikal and Lake Ohrid on the Macedonian-Albanian border, have been declared World Heritage Sites (Ohrid because it is one of Europe's deepest and oldest lakes — and most biodiverse, with more than two hundred endemic species). Five more lake basins have been declared biosphere reserves: Champlain in New York, the American Great Lakes, Issyk-Kul, Malawi-Nyasa, and Tonle Sap. These designations have helped publicize their advantages — and their problems.

It's a start.

Part Two

Issues

4
Who Owns Water?

All the great themes of modern water management go back to ancient times. And who is to say that the ancients were less adroit than we are? Technologically unsophisticated, yes, but politically and economically...not so much.

Water privatization and the "commodification" of water? More than twenty-four hundred years ago, Darius the Great of Persia, a ruler noted for his monumental building projects and superlative management skills, encouraged the construction of new water supplies by exempting from taxation "for five generations" the revenue derived from the building of new qanāts (a qanāt is a kind of horizontal well, bringing in water from uplands, sometimes a considerable distance away). Clearly, Darius wanted to get the thing done, and he did so in an early instance of the government pump priming the economy by subsidizing entrepreneurs — tax breaks for making a profit doing a public good.

Water justice? One of the major themes of modern water law is enshrined in the "do no harm" notion, in which upstream users can freely use water, but not at the expense of folk lower down the flow. We are still wrestling with how to manage this idea, even after multiple UN conventions and literally hundreds of water treaties. This too was governed by water law in ancient times. The lawgiver Hammurabi, who died somewhere around 1750 BCE, gave some thought to this notion. Article 55 of his legal code said this: "If anyone open his ditches to water his crop, but is careless, and the water floods the field of his neighbor, then he shall pay his neighbor corn for his loss." The next article, 56, provided penalties: "If a man let in the water, and the water overflow the plantation of his neighbor, he shall pay ten *gur* of corn for every ten *gan* of land." (A *gur* is around three hundred litres, and a *gan* is a square approximately 53 metres on a side.) Almost two thousand years later, the Talmud ruled that "no man may sell water from a public cistern and therefore profit from what the people have done."

Crop insurance and forgivable loans? Hammurabi's Article 48 said this: "If anyone owe a debt for a loan, and a storm prostrates the grain, or the harvest fail, or the grain does not grow for lack of water; in that year he need not give his creditor any grain, he washes his debt-tablet in water and pays no rent for this year."

Water as an investment? Solomon was there: "I made pools of water to irrigate a forest springing up with trees... Thus I gained more wealth than anyone else before me in Jerusalem."

The right to water? Maimonides' Book of Acquisition says this: "When people have fields along a river they water them in the order [of their proximity]. But if one of them wants to dam up the flow of the river so that his field may be watered first, and then reopen it, and another wants to water his field first, the stronger prevails. The cistern nearest to a water channel is filled first in the interests of peace."

The Quran insists that no man may abuse a well, and if a man has surplus water, he must provide it to strangers and their cattle (though not for irrigating crops).

The right of first use. The right of might.

Thus, the politics of water.

There are more of us now and the consequences of inaction are more dire. But the challenges are the same: how to manage fair distribution of a scarce resource, and how to ensure sustainability. Which means this: Who owns water? Who processes it? Who controls its availability and distribution? Who should? Who has it and who doesn't? Who wants to steal it? Who can and does?

Catherine Brölmann, in a smart blog for the website brokeronline. eu in 2013, put it this way: "In recent years international law has increasingly started to address these challenges... [but] a fundamentally communal interest (or as some would say, 'a global public good') such as water is not easily translated into legal terms. International law — famously lacking a world constitution or a central legislator — is a system traditionally based on freely agreed commitments by territorially sovereign states." It is all complicated, as she points out, by the fact that water and waterways have for centuries been viewed as part of sovereign territory, a doctrine that gained extra weight in the decolonization era, when new states jealously guarded their new prerogatives.

She also believes, though her evidence here is thinner, that in the last decade or so the stalemate over territoriality has "to some extent been overcome by a genuine paradigm shift: from water use as a state's right to water access as a human right." She admits that the UN resolutions pushing this attractive notion are not binding. And she points out that Canada, for example, has been "conspicuously reluctant" to recognize the human right to water just in case the parched southwestern United States ever thought of suggesting it was Canada's moral obligation to provide water where it had none and Canada plenty.[1]

International Water Law

Brölmann aside, there have been plenty of efforts to codify international behaviour into law.

An early effort was American, but it became influential because it was used to justify water actions in other jurisdictions. The so-called Harmon Doctrine from the middle of the nineteenth century was named after a decision by Judson Harmon, an American jurist, who was called to answer a Mexican complaint that Americans were taking too much water from the shared Rio Grande River. To Mexican annoyance, Harmon was dismissive: a state, he maintained, is free to use the waters of an international river that are within its own territory in any way it wants, without regard for the harm it might cause any other riparian state. (This would come back to haunt Texan farmers in later years, when the Mexicans abstracted large amounts of water from the part of the same river that flowed through Mexican territory.) Applied elsewhere, the doctrine also caused no end of trouble over excessive abstractions from the Jordan River, in the Middle East.

Little law was made in the early years of the twentieth century — no doubt a couple of world wars had something to do with it — but the post-colonial era brought a new focus on international water management. In 1966, after decades of wrangling, the UN and a slew of developed countries adopted what came to be called the Helsinki Rules. This was among the first documents to establish that drainage basins, and not nations, should be the operating unit for water management. It also codified a rule that more than a dozen cooperative bilateral treaties had already suggested: the principle of the "reasonable and equitable" use of available water. The Helsinki Rules laid out the criteria to be assessed: history of past use of the basin, the economic and social needs of all users, and the availability of alternate sources.

A decade later, in 1978, the UN's Food and Agriculture Organization (FAO) issued a catalogue of more than two thousand treaties and "international instruments" dealing in one way or another with international watercourses and aquifers, some of them dating back to the first or second centuries. Most of these agreements were bilateral, and dealt with shared boundaries or rivers that flowed from one country into another. They range from the Amazon Cooperation Treaty Organization adopted by eight countries, through the Lesotho Highlands Water Project (South Africa and Lesotho) and the Mekong River Commission, from which China is conspicuously absent, to the Zambezi River Authority (Zambia and Zimbabwe). It has been the task of the codifiers of law to try to cobble together a list of general principles from all these many agreements.

Two more decades of negotiations followed before the cumbersomely worded "UN Convention on the Law of the Non-Navigational Uses of International Watercourses" was agreed to by the UN General

Assembly in 1997 and finally adopted in 2014 — many countries got cold feet and ratifications were slow in coming.

Throughout these decades, efforts to codify water law were somewhat complicated by the fact that multiple bodies with varying degrees of influence were attempting to draft the rules. Only one is in any way "official," that is, bound by either national or international law, but two others carry considerable weight too. The official one is the International Law Commission (ILC); the other two are the Institute of International Law (IIL), and the International Law Association (ILA). The ILC is a creature of the UN; its thirty-four members are elected directly by the General Assembly and work on projects mandated by the UN. The IIL is a group of self-perpetuating and self-defined "internationally worthy jurists," founded in 1873, that has gained some traction in that its rulings have often been cited in diplomatic negotiations. The ILA, by a curious coincidence also founded in 1873, is another professional body, numbering around a thousand, drawn from a wide variety of countries; it too has no official backing but its recommendations on international law are also taken seriously.

All three have developed legal principles on water law, each with slightly different shades of meaning. In 1961, the IIL adopted what it called the Salzburg Resolution on the Use of International Non-Maritime Waters, and eighteen years later followed that with a further declaration, the 1979 Athens Resolution on the Pollution of Rivers and Lakes and International Law. The Salzburg Resolution set up the principle of equitable utilization, by which it meant that sovereign rights over international watercourses are limited by the "right of use" of other states sharing the same water. Disputes should be "settled on the basis of equity, taking into consideration the respective needs of the States, as well as any other circumstances relevant to a particular case." That's pretty loose: to make it work, it requires maximum flexibility, much discussion, and the setting up of bodies of experts to foster cooperation. Obduracy by any party generally scuttles agreements.

The ILA's Helsinki Rules on the Uses of the Water of International Rivers also supported the equitable use notion. States are entitled to a "reasonable and equitable share in the beneficial uses of the waters of an international drainage basin." The Helsinki Rules explicitly rejected the "first in time, first in right" and the "use it or lose it" principles. For the first time, controversially, the Helsinki Rules declared that an existing use may have to give way to a new use in order to come up with an equitable distribution. There was provision for compensation, but "first come, first served" would no longer apply.

The ILC, for its part, maintained that while equitable utilization was all very well, a state had no right to *any* use if such use harmed another riparian state. "A watercourse state may not justify a use that causes appreciable harm to another watercourse state on the ground that the use is 'equitable' in the absence of agreement between the states concerned." In Article 7 of the draft of the Convention on the Law of the Non-Navigational Uses of International Watercourses, the commission was even more explicit: "*Prima facie* at least, utilization of an international watercourse is not equitable if it causes other states harm." That was a bit much for a large number of states that were concerned their prior rights would slip away if this notion of downstream harm were to be enshrined. The ILC backed off in 1994, and equitable distribution became the principle embedded in the UN convention. The Helsinki Rules have been widely adopted — for example, the International Court of Justice used the Helsinki framework when it ruled on a dispute between Hungary and Slovakia.[2]

International efforts to codify water law didn't stop with Helsinki; virtually every water conference now includes a session on "Water and Law." As the December 2000 issue of *Water International* put it, lawyers and politicians are still wrestling with four key principles:

- Legal entitlement (What is the scope of the resource and who is entitled to it?)
- Framework for allocation (Where all needs cannot be met, who is entitled to what quantity or quality of the resource?)

- Institutional mechanisms, including governance issues (Who is responsible for implementing, or overseeing the implementation of, the laws?)
- Compliance verification, dispute avoidance and resolution (How are rights and obligations enforced?)

In the year 2000, four years after the adoption of the Helsinki Rules, the EU council approved the Water Framework Directive, described in Chapter 3, which had as its backdrop that "water was not a product like any other but a heritage which must be protected, defended, and treated as such." The directive was mostly aimed at preventing abuses, rather than allocating rights.

In 2004, the ILA summoned another international conclave, this time in Berlin. The Berlin Rules followed, and in 2014 were still the last word on the subject, though not everyone has agreed to use them.

The Berlin conference was the consequence of growing uneasiness among water policy experts everywhere about environmental issues, the kind of issues the EU Water Directive was meant to capture. A new awareness of "the need to protect the integrity of the aquatic environment," as the report from the conference put it in its preface, along with "the prospect of global climate change [that] could worsen the situation dramatically," were testimony to both the newly increased influence of the environmental movement and the inroads climate scientists were making in overcoming policy inertia.

Again from the Berlin preface: "The Helsinki Rules...played a key role in formulating the rule of equitable and reasonable utilization as the basic rule of international law for the trans boundary use and development of waters. Today, the situation is significantly different than in 1966. The steady decline in the water available *per capita* in much of the world by itself poses serious challenge to international law as it had developed until 1966. Moreover, the emergence of a body of international law addressing significant environmental problems and the protection of basic human rights since 1966 has broadened and deepened the role of customary international law applicable to the world's waters. The International Law Association approved supplemental rules from time to time to respond to some of these new realities."

The differences are mostly shadings of what went before. What was new was the repeated emphases on environmental stewardship. Water as a human right, a key tenet of the social justice movement, gets an approving nod in Rule 17: "Every individual has a right of access to sufficient, safe, acceptable, physically accessible, and affordable water to meet that individual's vital human needs; States shall ensure the implementation of the right of access to water on a non-discriminatory basis."

What was new, also, was a worry about war and conflict. A number of rules dealt with military action. For example, "Combatants shall not poison or render otherwise unfit for human consumption water indispensable for the health and survival of the civilian population"; "Combatants must not target ecological targets"; "Combatants must not target dams and dikes"; and "An occupying State shall administer water resources in an occupied territory in a way that ensures the sustainable use of the water resources and that minimizes environmental harm . . . and ensure an adequate water supply to the population of an occupied territory."

This was written before rebels in Iraq seized the country's second biggest dam, near Mosul, and before it was taken back again and become a continual tug of war, literally. The dam itself has not been destroyed. Not yet, anyway.

In more recent years, two other concepts have begun to appear in water literature and at water conferences. The first, a product of the same forces that are spearheading the anti-dam movement, is called "integrative management strategy," really water-as-a-human-right from another angle. This strategy posits that international agreements should recognize fairness, equity, and justice, and that policy should "focus on marginalized or vulnerable populations and consider issues of social and political power in decision making." While worthy, this has scant chance of going anywhere.[3]

The second concept grows out of the perceived need to treat water management (in the jargon of our times) "holistically." This is called IWRM, for Integrated Water Resources Management. As Maggie Black and Jannet King put it in their *Atlas of Water*, "Among other things, this emphasized the river basin as the logical geographical unit for strategic planning, making cooperation between basin partners even more critical. The approach also encompasses... environmental protection, food security, especially for the poor, appropriate choices for water use in economic productivity, good governance including decentralization of decision making, reform of water managing institutions, effective regulation, cost recovery and equitable pricing."[4] This one does have traction, though John Briscoe, now a Harvard professor but then a senior water manager with the World Bank who had something to do with developing this beguiling notion, believes it has got out of hand. As he said in a conversation in 2014, "I regret my involvement in elevating this idea. It started as a reasonable one, in which decisions had to consider externalities, but has evolved into a useless and paralyzing idea of 'everything is connected to everything else.'"

A World Water Council report in the year 2000 summarized where matters stood, in the deadening language of such bodies. What the world needs, the report said, was "holistic, systemic approaches based in integrated water resource management; participatory institution mechanisms; full-cost pricing of water services, with targeted subsides for the poor; institutional, technological, and financial innovation; governments as enablers, providing effective and transparent regulatory framework for private action; the mobilization of political will; behavioral change by all."

National Water Laws

It is more difficult to summarize national water law than it is international water law. Dozens of frameworks exist, even within countries. But it is fair to say that in most countries of the world, water "belongs" to the state that can allocate it, mediate disputes about it, license it, sell rights to it, forbid its use, and everything in between. It is also fair to say that in most countries — or at least in most countries where the notion of private property is acknowledged and protected — the water beneath a piece of real property belongs to the property owner. Thus, Libya can take as much water from the Nubian Sandstone aquifer as it likes without consulting anyone else, though the aquifer spills over several national frontiers. And Indian farmers can sink boreholes in the land beneath their crops and extract as much water as their pumps can bring up. Irrigators over the Ogallala in the United States can use powerful pumps to draw down the aquifer with nothing but their own self-interest to say them nay, at least until recently. There are also dozens of exceptions. One of them, I learned to my surprise, is in a majority of Canadian provinces, where water is "owned" by the province, which could, if it wanted to — and if it dared — forbid me to use my little well to deplete my little aquifer. Also in Canada, even more strangely, citizens have no legal right to clean drinking water, nor the government an obligation to provide it.

Surface water is more complicated than groundwater. Because it is more visible, jealous regard over rights and their infringement is more probable, and perceived infringements are legion. So regulation is necessary and universal.

Again, though, systems vary. In Canada, all "navigable waters" are vested in the Crown (Canadian shorthand for the state), in contrast to English (and most European) law, where something called riparian doctrine became the essence of common law with regard to water. The riparian doctrine just means that anyone whose land fronts a body of water has the right to use that water. They didn't own the water — that was part of the *res communes*, property that belonged to no one and therefore belonged to all — but they could freely use it. Through most of the second millennium in Europe, where populations were still low and water plentiful, that right persisted unchecked.

It is in the United States that water laws become most complicated, tangled up as they are with states' rights, an assertion of individual over communal rights, and a strong suspicion of federal regulation.

In general, there are two systems of American water law. East of the Mississippi, where water was generally plentiful, a version of the European riparian system prevailed — that is, you get a certain amount of water according to the size of your adjacent land holdings. This also meant that water was tied to land. It could not be bought and sold as a commodity. You didn't own it. You only had a "usufruct" to the stream or lake — a right of use, not of ownership. In most states too, usufructuary rights were not absolute; downstream rights holders had the right to water "in natural volume and unimpaired in quality." A commonly cited precedent was a Maine Supreme Court decision of 1845: "Every owner of land through which a natural stream of water flows has a usufruct in the stream, as it passes along, and has an equal right with those above and below him to the natural flow of the water in its accustomed channel, without unreasonable detention or diminution in quantity or quality, and to the reasonable use of the stream for every beneficial purpose to which it can be applied, and none can make any use of its prejudicial to the other owners, unless he has acquired a right to do so by license, grant, or prescription."[5]

In the second decade of the twenty-first century, as prolonged drought took hold in many parts of the United States, the long-discredited Harmon Doctrine once again peeped over the barricades. After several complaints from neighbouring states, a federal ruling was made: Georgia has been taking too much water out of the ACF river system, the drainage basin of the Apalachicola, Chattahoochee, and Flint Rivers, thereby harming people downstream. Georgia's indignant governor invoked Harmon and declared that Georgia was entitled to use all the water that originates in or falls on the state, with no responsibility to other users. "The state of Georgia is due the use of that water, and we will make use of that water," and Florida and Alabama be damned. Of course, Governor Sonny Purdue positioned the federal ruling as a case of intolerable bureaucratic overreach and, reflexively, blamed President Obama, but he was really harking back to Harmon. All this drew a post on the San Francisco website SFGate from Peter Gleick, the water analyst, who as usual had something sensible to say:

It is time for Georgia, or China, or any water user in places where water is no longer abundant to stop posturing and start discussing how to share our rivers and our ground-water too. Just because you are first on a river, or upstream, or own a piece of land with groundwater does not mean that you have no responsibilities to other users, including non-human users, sharing the same watershed. Your use affects others. We may believe that these water resources are not connected to each other or that the use by one person has no effect on their neighbors or that the first user should have senior rights forever. And in the past this may have worked. But if we continue to use 20th century rules to solve 21st century water problems, conflicts over water will only worsen.[6]

West of the Mississippi (and in the state of Mississippi itself), a different system prevailed. The West was settled by homesteaders, and scarcity was an issue from the start; the newcomers settled along what water-ways they found. So the basic law of the West became first come, first served, or, as the common law expressed it, first in time, first in right. In times of shortages, therefore, those with the oldest rights could go on using their water long after "junior users" had been told to stop. Also, rights to water were separated from ownership of land, which meant water rights could be accumulated — and sold. That was new.

Some limitations were imposed. The right of prior appropriation generally insisted that the user put the water to beneficial use, other-wise the rights could be lost — this is the infamous "use it or lose it" doctrine, perhaps the most anti-conservationist ethic of them all.

In Texas law, to take just one American example among many, water rights depend on whether the water is underground or above it. Landowners don't own the water underground, but they have the right of capture, that is, they can extract and use as much as they like, even if such use harms a neighbour. Not surprisingly, this has led to a good deal of litigation and is becoming more urgent as groundwater over-pumping becomes more of an issue. In 2014, the Texas Supreme Court upheld an absolute right of capture but pointedly suggested that the state might want to reconsider the relevant law. The state declined, and punted to local management boards.

An interesting case is the Edwards aquifer, rapidly being depleted by over-pumping. Farmers use it for irrigation; the city of San Antonio uses it for its one million plus citizens and some of its springs drain into the Guadalupe River, whose downstream users are demanding that their share be maintained. In 2011, the Edwards Aquifer Authority asked its users, including San Antonio, to reduce water consumption by 35 percent. The city complied. Not so pecan farmers Glenn and JoLynn Bragg, who sued, saying the state had violated their property rights. The courts agreed.[7] The aquifer is still emptying. As a curious sidelight, the Environmental Defense Fund, a staunch supporter of the commons but also in favour of "ecological capitalism," has suggested that the Edwards aquifer be privatized and that groundwater rights could then be quantified based on average use over decades — no actual amounts would be specified, only a share of the aquifer's safe yield.[8]

Surface water in Texas, on the other hand, belongs to the state, and private users need the state's permission to capture it. This has caused some interesting problems. The Texas oilman T. Boone Pickens, for one, has been buying up land so he can hoard the water beneath it with a view to then selling it to thirsty states. But how to get it there? If he brings it to the surface and doesn't use it himself, it will then run away into a river, or, if he builds a canal to take it somewhere specific, it is then surface water — that he can no longer own. (This being Texas and money being what it is, Pickens solved this pickle by getting the Texas legislature to declare the "Pickens exemption.") Even so, he has largely abandoned his notion for a massive pipeline to take water "everywhere"; instead, he's planning to move it mostly around the neighbourhood whence it comes, the Texas Panhandle.

Arizona is the only US state to govern or control private access to groundwater. Early in the century, the state's comprehensive groundwater law mandated sustainability by 2025, a deadline now being hurriedly reconsidered in light of the unremitting drought. The preferred method of balancing withdrawals and extractions, until recently, was the innocent notion of simply importing more water from the Colorado River. This too will no longer do.

California, where water is suddenly a much more critical issue than in past decades, long ago adopted a hybrid system called "correlative rights." This meant that landowners could freely use the water beneath them, but only subject to the reasonable use rule. A surplus, if any, could then be claimed by other landowners under the prior appropriation rule. If there is no surplus, the other landowners get nothing. It has led to some curiosities. While most pumping is to fill taps or keep vines and almond trees alive, a group of farmers in Merced County began pumping their own groundwater and then selling it. The *Modesto Bee* newspaper reported a $46 million, four-year deal to sell nearly 100,000 acre feet (somewhere around 123 million cubic metres) to buyers in the neighbouring county. Outraged local authorities promptly scuttled the deal.[9]

This all changed in 2014. In August, the California legislature abruptly passed new and sweeping groundwater rules. Landowners could no longer count on an absolute right to the water beneath their feet. The new measures didn't abolish private ownership exactly, but future withdrawals would have to be done through and with the say-so of local water agencies. And if the local agencies failed to come up with a comprehensive management plan, the state could step in and take over. It was a measure of the severity of the California drought that few squawked.[10]

Finally, it is worth mentioning Australia. Australia, another generally arid place — it is certainly the driest continent on earth — followed British common law until the 1980s, when it adopted the western American notion of allowing the accumulation (and therefore the trading) of water rights. At the turn of the twenty-first century, partly in response to a series of prolonged droughts, the Australians went further, even setting up brokerages and exchanges to make water trading easier. Not only riparians but investment firms, even foreign ones, can now own water rights, and buy and sell them freely. As a result, Australia is currently the world leader in marketing water. We shall see, in the next chapter, how this is working.

5
Privatization and Its Discontents

Some stories seduce by their very improbability — a story so unlikely is unlikely to have been made up, therefore there must be something to it — thus the conspiratorial logic of the Internet. One such story was circulating as recently as 2012, and it involved "the Bush girls," the daughters of former president George W. Bush. Barbara and Jenna Bush, the daughters in question, were reported to be using family money to buy up aquifers in South America, their purpose opaque but surely aimed at making a profit on the backs of... well, whoever owned the water in the first place.

None of this was true, but you can see where it came from. The notion dates back to a 2007 "investigative report" claiming that George W. had bought some forty thousand acres in Paraguay. The story was then floated onto the wilder shores of the Internet by wishful thinkers who assumed he was looking for a political bolt hole in a country that did not have an extradition treaty with the United States, to avoid prosecution for war crimes in Iraq and elsewhere. Later it was noted that this supposed purchase (said to be assisted by cult religious leader Sun Myung Moon, just to layer on another level of improbability) lay atop part of the South American Guarani aquifer, a massive reservoir introduced in Chapter 2. From there, of course, it became "obvious" that the family was intent on profiteering at the expense of the indigenous poor.

As I recounted in Chapter 2, on groundwater, the Guarani has for some reason attracted numberless crank theories in the last decade or so, but even by those standards, this is pretty wild stuff. However...is it any wilder than the assertion that a powerful water cartel, sometimes described as a "clique," has emerged to seize control of the world's water for its own profit? Listen to this: "The rich will drink only bottled water found in the last few uncontaminated parts of the world, or sucked from the clouds by corporate controlled machines, while the poor will die in increasing numbers from a lack of water. This is not science fiction. This is where the world is headed unless we change course — a moral and ecological imperative." This is from the Canadian polemicist Maude Barlow.[1]

Or this, from Joanna Robinson, author of a mostly sensible anti-privatization book called *Contested Water*:

> The politics of water and the initiatives of social movements fighting to ensure protection of and fair access to water will be among the most important in human history....Many activists argue that water is the next oil, and that the wars of the future will be fought over control and access to water. The battles have already begun, as communities around the world, from Cochabamba, Bolivia, to Atlanta, Georgia, and Stockton, California, have wrestled with the question of who should control their water. A matter of life and death, the politics of water, including the movements that mobilize to protect water as part of the commons, is one of the most critical, visible and contested issues of our times.

On the other side of this debate are the think-tankers of the right, for whom the private markets can do no harm. To take a random example, here's a passage from Fredrik Segerfeldt, whose book, *Water for Sale*, was Englished from the original Swedish by the always reliably anti-government Cato Institute in the United States: "Concerning water as a raw material, a completely free market with tradable water rights will result in availability increasing, prices falling, and the water going where it does the most good." There is no discussion of this assertion, just the bald assertion itself, as though it were a truism that needed only nods of agreement. But it is wise to be skeptical when an author declares in his preface: "In this book, then, we will be leaving dogmatism and ideology aside in order to discuss why water distribution in poor countries is in such a wretched state, what has been done, and what can be done."[2]

Polemics aside, it *is* true that a dozen or more financial indexes now exist that list the stocks of private corporations that have been investing in water rights and infrastructure — spending water like money, as the saying goes on the street. It *is* true that there are mutual funds now entirely invested in water stocks. It *is* true, as recounted earlier, that the billionaire tycoon T. Boone Pickens has been buying up rangeland in Texas and other places for the groundwater beneath it, with the declared intention of peddling it at a markup to whatever buyers he can find. (It is *not* true, as per the Bush girls legend noted above, that Pickens is planning to add an antipsychotic drug to the drinking water of every American. At least, I don't *think* it is…) It *is* true that at least one large hedge fund has decided infrastructure investing is boring and is buying up what its manager is pleased to call "wet water," the actual stuff as opposed to the means of delivering it, though this is a bit of a verbal swindle: what it is really buying are rights to use it, not the stuff itself, and rights like this are available only in a handful of jurisdictions. It *is* true, in those few places where it is legal, that those rights can be bought, hoarded, or resold to anyone, anywhere — in Singapore, Zanzibar, or anywhere in between. Of course, the water itself doesn't go anywhere — you can't move it to its new owners in Singapore or Zanzibar if the actual stuff is in Texas, except at ruinous cost.

In the last few years, early investor enthusiasm has begun to cool off, but much of the current rhetoric hasn't caught up. It is still excitable, having switched from "The wars of the 21st century will be fought over water," a phrase popularized by World Bank vice-president

Ismail Serageldin in the 1990s, to "Water is the new gold," attributed to a *Fortune* headline but now so common as to be public currency. It is precisely this kind of assertion, with its underlying sense of smug entitlement, that grates on the sensibilities of the anti-privatization activists. Underpinning all this is the increasing commodification of water, at least partly spurred by the very water overdrafting the UN has identified: in classical economic theory, after all, scarcity equals increasing value and profit opportunity — so water *should* be profitable, no? The newsletter *MarketWatch* gleefully points out that, in 2010, global water generated somewhere north of half a trillion dollars of revenue, a number expected to swell by more than 40 percent by 2030, as population, and therefore demand, increases[3]....Or so say the analysts.

It is really not so simple. And "global water" in this formulation is a lot more slippery of definition than it seems. Nor is the market as "hot" as promised. By some measures, the "private water behemoths" of the anti-globalist polemicists are no longer in rapacious mode, if they ever were, but instead are in more or less unruly retreat. Some polemicists assert that the commodification of water is a wholly beneficial thing, others that it is just another manifestation of the neo-liberal assault on the commons and the common good.

And, of course, maybe it is both. Perspective is everything.

The Cochabamba Effect

The most famous name in anti-privatization circles is Cochabamba, a Bolivian city of around a million citizens in a reasonably verdant Andean valley, reasonably well watered by glacial melt, surrounded by farmland devoted mostly (until the American war on drugs) to local foods and coca plantations. Pro- and anti-privatization interests concede that the Cochabamba debacle represents the most dramatic failure to date of efforts to improve water supplies through private corporations. It is not the only failure, but it was one of the first and most dramatic, and it emboldened the opposition in many other parts of the world. The whole thing happened quickly. Water privatization in Bolivia was enabled by a law passed in 1998. A mere year later, a forty-year concession was granted to Aguas del Tunari (AdT), a local firm partially owned by the American engineering company Bechtel, based in San Francisco (usually muscled-up in anti-privatization prose as "the American water giant Bechtel," which is a little misleading: Bechtel is a huge engineering company — it helped build the Hoover dam during the Depression — but water isn't its main business). Less than a year after that, the company was gone, driven out in a massive public uprising fomented by a coalition of peasant farmers, the urban poor, traditional water vendors, and even sections of the urban elite. This loose-knit group was joined by the highly organized federation of coco-leaf growers, led by the charismatic president-to-be Evo Morales, who brought with him considerable experience in organizing street demonstrations, road blockages, strikes, and general civic disruption.

All of which followed: street protests led to ever more massive demonstrations, followed by a violent response from the military, followed by widespread work stoppages and overall social mayhem. During the rioting, dozens were injured and three demonstrators were killed; the disruption was bad enough that the business-friendly president of Bolivia, Gonzalo Sánchez de Lozada, was subsequently driven out of office, as was his neo-liberal successor, opening the way for the election of Morales. Control of the city's water was then returned to the municipal utility, Semapa, from which it had been wrested.

So much, both sides can agree on. But there are many devils in the details. The anti-privatization case is this:

Semapa was conceded to Bechtel in exchange for debt relief for the Bolivian government and the promise of new World Bank loans to expand the existing water distribution system. This was an expression of the neo-liberal vision of governance, in which private expertise was always preferred to democratic controls. The contract of concession gave Bechtel an effective forty-year monopoly over water supply in the Cochabamba Valley. Not just over rivers and groundwater, either: the monopoly included private wells on private property, backyard rainwater cisterns, private extractions from creeks and rivers, and a network of water vendors that had been delivering water by truck to poor communities unserved by the utility, as well as the dozens of existing community-based village and rural water co-ops in peri-urban areas around the city. "Even the rain," it was said at the height of the subsequent protests, "would be owed to Bechtel." The quid pro quo was that the corporation was supposed to improve and extend the piped-water grid to places it had never reached before. Instead, Bechtel immediately tripled the price of water, cutting off those who could not pay. In a country where the minimum wage is less than $60 a month, many users received water bills of $20, which they could not afford. The company even charged them for rainwater collected in cisterns. The Bolivian revolt is therefore positioned as a consumer rebellion against water privatization, fuelled by price gouging and indignation at the anti-democratic nature of foreign corporations controlling access to an essential service. In the aftermath, the city's water was successfully returned to public control. The coalition of social movements that had precipitated Bechtel's ouster, called the Coordinadora del Agua y de la Vida (the Coalition for Water and Life), set out to transform Semapa into a truly democratic public service provider, run under the principles of "social control," that would give the poorest, who generally had no water or sewage, a voice in how the utility should be run.

Well, it is a nice narrative. But is it true?

By way of background, consider the state of affairs prior to Bechtel. Semapa had been set up in 1967 to bring a semblance of order to the shambolic and uncoordinated ad hoc coalitions of private vendors and small water co-ops that had hitherto delivered what water they could. Thirty years later, Semapa was itself in shambolic condition. It had nearly five employees per thousand connections, double the rate of a well-run utility. Much of what meagre revenue was generated often disappeared through graft, some of it going to city hall and to mayoral cronies. Less than 57 percent of the population was on the municipal water grid at all, and the proportion was going down as the population swelled. Half of all water was lost to leakages in rotten pipes. The poorest people, whose connection rate to the grid was less than 5 percent, paid the most for water, being entirely dependent on a fleet of tanker trucks that showed up in their neighbourhoods on a weekly schedule and charged whatever they could get. A tenth of all connections to the grid were illegal, but scofflaws went unpunished. Many of the larger customers, mostly public sector companies, refused to pay anything for their water, and Semapa lacked political muscle to force compliance. Unsatisfied demand was estimated at nearly 40 percent. In many areas of the city, water was available only for a few hours a week, and rationing was imposed inadvertently: the taps just didn't run when there wasn't enough water. And so on.

Part of the plan to fix all this, to rationalize the chaos, was to bring in outsiders with experience — and money. Another part, proposed by the city's mayor, Manfred Reyes Villa, was to build a large 120-metre-high storage dam on the nearby Misicuni River that would include a nineteen-kilometre tunnel and a hydroelectric plant. To get it started, a new public sector corporation called Empresa Misicuni was set up, consisting of Semapa and the departmental and municipal governments of Cochabamba, with a minority stake held by the national government. The driving force behind all this was the mayor; it was he who insisted on the dam, partly under pressure from some of Bolivia's biggest and politically connected construction companies, which clearly expected lucrative contracts to build it.

It was at this point and under these circumstances that the Bolivian government, strapped for cash, turned to the World Bank and its advisers. But with unexpected results: a World Bank feasibility study concluded that while extending the grid was desirable, the dam would cost far too much (somewhere around $300 million) and could not be justified by its expected results. The bank therefore declined financing. The mayor then went it alone and sought out private-sector partners. The concession contract with Bechtel, bid without competition and without external oversight, resulted.

It didn't start well. Under the contract, AdT, the firm partially owned by Bechtel, was indeed given exclusive use of water resources in the valley, as well as "any future sources" needed to supply the city, and was granted exclusive rights to provide water services and to require — a fraught word — potential customers to get hooked up. (But no, the rain wasn't included, nor were backyard cisterns.) This made enemies immediately, riding roughshod as it did over the informal network of well drillers, small water co-ops, and truck-based water vendors that had been operating to that point. In most privatization concessions, a grant of exclusivity is reasonable, as water delivery is generally a natural monopoly and competition is generally absent. In Cochabamba, though, competition did exist. It was small but it was numerous, and it became a real force in the protests that followed.

To cover the costs of extending the grid and fixing its many deficiencies, the contract stipulated an immediate tariff increase of 35 percent, plus an additional 20 percent by 2002, when, it was believed, water should start flowing from the Misicuni dam. But this 35 percent, which soon turned into 43 percent, wasn't increased across the board. Instead, it came through what the contract called an "increasing block tariff," in which residential consumers were divided into categories that included "precarious constructions" (slums), "economic dwellings and functional apartments," and luxury housing. High-income householders would pay three times as much as low-income householders for the first twelve cubic metres, and twice as much thereafter.

Still, while the tariff was hardly triple, prices did go up sharply enough to cause resentment — and under the contract, AdT was guaranteed a 16 percent return on investment, a ridiculous rate and itself a source of considerable outrage.

On the very poorest consumers, prices went up between 10 and 15 percent, which meant they were on average paying between 6 and 10 percent of their income on water — high but less than what they were paying water vendors in the earlier private market. On higher income consumers (some of them still not very well-off), prices increased as much as 106 percent. On certain others, they increased more than 200 percent — these were generally companies that had fiddled the categories under the old system, calling themselves households in order to qualify for a lower rate, and were now re-categorized, much to their indignation.

The popular rebellion that ensued wasn't just about water. The conflict took place at a time of continuing social unrest in Bolivia, grave enough for a state of emergency to be declared in 2000. In the wider context, the government's efforts to restructure the economy, which in practice meant the capitalization of state-owned enterprises and continuing privatization of services, brought in much less revenue than forecast, and economic growth sagged. By the year 2000, as much as 70 percent of the population was defined as poor. In addition, there was a long and quite successful tradition in Bolivia of cooperatives and local management boards, which the new system traduced.

The Cochabamba "water war," therefore, was of a part with the larger protests that identified the country's multiple problems with the neo-liberal development strategy, compounded by the angry protest by the *cocaleros*, or coca farmers. The final element was Bechtel/AdT's determination to eliminate the traditional non-state water suppliers, well drillers, and tanker trucks, all of whom promptly joined the protests. All this was further complicated by an unexpected success on AdT's part: a fall in leakage rates led to a reduction in the need for water rationing, and new hookups encouraged many consumers, even the poor, to rapidly increase their consumption. This, of course, put their bills up.

More than a decade after Bechtel ignominiously departed ("fled" is not too strong a word), what is the outcome? Some progress has been made, but the poor are still paying more for their water than the rich, not yet being on the grid. Water is available for only a few hours a day, its quality is deteriorating because too many poorly sealed wells are contaminating groundwater, and the Misicuni dam has not yet been built — tenders went out in 2009, were rescinded, and offered again in 2014. The buffer its water would have offered, and its attendant hydro power, remain a pipe dream. The management company, once again a public body, is still inept and inefficient. The mayor's office and the

BACK TO THE WELL

union still treat revenues as a convenient opportunity for enrichment. A third of the water is still lost to leakages and clandestine hookups. As Emily Achtenberg wrote, "In 2010, the public water authority was forced to lay off 150 workers, because it found itself with a $3 million cash deficit, due to alleged irregularities such as payroll padding, materials theft, and continued diversion of the system's water."[4]

But there were some real gains. A second water war, this time in El Alto, the poorer outskirts of La Paz, bustled another private water giant out of the country, this time Suez, of France, which had bungled its contract with the La Paz–El Alto water district. In 2003, another revolt, inspired by Cochabamba, forced the cancellation of a plan to sell off natural gas through a pipeline to Chile at rock-bottom prices. Bolivia's new constitution, enacted in 2009, proclaims that access to water is a human right and bans its privatization, only the second country to do so, after Uruguay. Externally, Cochabamba encouraged and inspired anti-globalists everywhere. At the UN, Bolivia led the successful campaign to recognize water and sanitation as a basic human right.

None of this is yet getting much water to the poor of Cochabamba, who are as water-deprived as they ever were. Ten years after Bechtel fled, the renationalized water management had yet to hire a single water technician.

The lesson is not that privatization is bad, but that bad privatization is bad. Cochabamba was a bad contract and deserved to end, but the manner of its ending made everything worse. "Kicking out a foreign corporation hell-bent on profiting handsomely off water was a major victory," as water activist Jim Shultz put it.[5] But it wasn't. It just traded one bad system for another.

On the upside, Evo Morales was re-elected president of Bolivia again in 2014 because his government was actually seen to be working pretty well. GDP has grown steadily under his presidency, and Bolivia has seen the highest rate of poverty reduction in Latin America, admittedly from a high base. UNESCO declared Bolivia free of illiteracy after a massive infusion of money into education. Morales's socialist-style nationalization of key industries, which the neo-liberals confidently expected would lead his country to ruin and to the exodus of major companies, had no such effect — the companies stayed, and the country prospered, more or less. As Benjamin Dangl put it in an Al Jazeera commentary, exaggerating just a tad, "Consider the fact that before the 1952 National Revolution, indigenous people weren't even allowed to enter the Plaza Murillo in front of the presidential palace because they were believed to be too dirty and unsanitary. Now an indigenous president and poor farmer without a college education sits in the presidential palace itself."[6] And, he might have added, is doing rather well. Cochabamba was, even if indirectly, at least responsible for that.

What Exactly Are You Privatizing When You Privatize Water?

Are we yielding control of our most precious resource to ... to *someone*? Or *something*? Are we, as opponents of privatization fear, in the process of allowing for-profit companies to own the world's water and resell it for profit? Is it true, as the American journalist and irrigation crank William Ellsworth Smythe once colourfully suggested, that "next to bottling the air and sunshine no monopoly of natural resources would be fraught with more possibilities of abuse than the attempt to make merchandise of water?"[7] Or is it a much lesser affair — are we just assigning ownership to water delivery or contracting out management? If so, to what purpose and to what effect?

Water, as we have seen, is elusive. As the American legal scholar William Blackstone once wrote, "Water is a moving, wandering, thing, and must of necessity continue to be common by the law of nature; so that I can have only a temporary, transient ... property therein." It moves, runs away to the sea. It is absorbed. It vanishes into the earth. It falls down as rain. Your body transforms it as it runs through you. How do you own such a thing? And how can you privatize something that is not ownable?

In fact, privatization alarmists aside, there are very few systems, anywhere, with completely privatized water assets and completely deregulated supply. Almost always, "privatization" means various degrees of partnership between public and private sectors.

A useful way of looking at the issue is to differentiate between ownership and use; a common saying in the water world is that you can own a bottle of water, but after you drink it, you don't own the water anymore, only the bottle. So the real question is not about ownership, it is about who is allocated the right to use the water that passes through. John Briscoe, the Harvard professor I quoted earlier, puts it this way: "There is no logical [or necessary] connection between the ownership of the resource and the way in which the service is provided. You can have any combination of... bulk water right, on the one hand, [and any] form of provision of services, on the other."[8]

So there are multiple ways in which water and its uses can be structured. In the United Kingdom, rights to both water and its delivery are owned privately by regulated quasi-monopolies with tight performance controls. In France, to take another rich-world example, there are no private rights to water, but in many cases, water services are provided by private companies under concession contracts. The Australian case is almost the reverse. There, farmers, cities, and corporations have private, tradable, "owned" rights to water, but the actual services are still provided by public utilities. Even in Australia and the western United States, what are traded are usufructuary rights, not ownership.

Another point is to recognize water's special value. It can be treated as a commodity and priced accordingly. But it is not like aluminum, say, or gold, which can be extracted from the earth, and then moved and repurposed effectively. You can't do that with water, except under limited circumstances. There is no economically feasible way to get the water from, say, Oregon to Yemen, even if you have a willing seller and a willing buyer. Only watersheds and river basins, and to a more limited extent aquifers, make water commerce possible.

Most of the confusion in the privatization debate, and much of the heat generated, is because the two separate issues of the resource and its delivery for use have been conflated. When we talk about privatization, we are almost always talking about infrastructure privatization. Three main models exist, each with different shadings.

The first is to contract out the operation and maintenance of existing services, with ownership still vesting in the original utility. Operation in this sense could mean not just running the facility but meter reading and bill collection, purification services, and so on, while maintenance could mean repairing and extending existing networks and providing new hookups.

The second, generally used where distribution systems either don't yet exist or are in such parlous shape that they might just as well not exist, is called DBO, which stands for "design, build, and operate." This calls for a comprehensive operating agreement and is also where anti-privatization activists get their ammunition, because many such contracts have proven poor value to the community, very often exempting the private contractor from the cost of fixing a system that goes wrong. There are dozens of examples, and we'll meet a few of them in due course.

The third is a simple asset sale. That means selling all community-owned assets in water and wastewater to a private contractor, albeit with maintenance and safety provisions built in. Rich-world countries hardly ever do this, though in the United States some small water utilities have been sold.

There are shadings in all three of these models. The most basic version is a limited-service contract in which the private contractor is hired to maintain existing networks or run the distribution system, which remains public property. Or a contractor can build and operate a system and then lease it for a fixed term. Or, under a concession contract, a private company can rent available infrastructure but undertakes to achieve a set of fixed targets, such as extending the network, maintaining pricing, and monitoring quality. That is, concessions grant a licence to run but not own the system, usually for a fixed term. This is by far the most commonly employed privatization.

You can also privatize other aspects of the hydrological system. Private companies can build desalination plants, for example. They can build pipelines and aqueducts, perform laboratory work, install purification systems, and accumulate grey water and stormwater, cleaning it and returning it for reuse. They can bulk-transfer water from one location to another (or at least they can try — so far no one has found this in the least practical, though many entrepreneurs are having a go). They can bottle ordinary water and sell it at outlandish prices. They can even build dams, or design and build turbines. These and other activities like them account for much of the market capitalization of water companies on the world's stock markets, and most of them are also not very controversial. What matters is not whether a dam builder makes a profit, or whether a pipefitter can price its wares for long-term profitability. What people care about is getting plentiful, clean, safe water to ordinary people at a price they can afford.

It is certainly possible to price water too cheaply and thereby encourage its profligate use — this is, alas, too often the norm in the rich world. It is also certainly possible to price it out of reach of the poor. That's what the privatization debate is really about, in the end.

Big Water

Hype aside, how big *is* the private and corporate water business?

If you are going to include just the international water companies (like Suez and Veolia Environnement and Thames Water and Trent Severn and Bechtel) that have contracted to run water and wastewater utilities in countries around the world, and the dozens of service and ancillary industries that have sprung up around them, a popular analyst's estimate of the market, based on no particular evidence that I could find, is that it represents somewhere between $400 billion and $500 billion annually. This is probably conservative. If you are going to include a host of extraneous matters that are related to global water, such as infrastructure repair and creation, or virtual water, desalination plants, and even household filters and purification systems, meters, and gauges and bottled water, there is no realistic way of totting it up and arriving at a global total, except to say that it would run into the trillions.

A trillion is a nice round number. It is probably the number that got the stock markets and their business-press toadies so fluttery in the early part of this century. That, and the heady notion that with water you have got, at last, a product for which demand will never shrink. One typically extravagant extrapolation came from an economist, Willem Buiter of Citigroup, who should have known better: "I expect to see in the near future a massive expansion of investment in the water sector, including the production of fresh, clean water from other sources (desalination, purification), storage, shipping and transportation of water. I expect to see pipeline networks that exceed the capacity of those for oil and gas today. I see fleets of water tankers ... Water as an asset class will, in my view, become eventually the single most important physical commodity-based asset class, dwarfing oil and agricultural commodities."[9]

Here's the text of the *MarketWatch* newsletter I mentioned at the start of this chapter. Its language is altogether typical:

> In 2010 global water generated over a half trillion dollars of revenue. Global world population will explode from 7 billion today to 10 billion by 2050, predicts the United Nations. And over one billion "lack access to clean drinking water."
>
> Climate and weather patterns are changing natural water patterns. And industrial pollution is making water a scarce commodity. So the good news is that huge opportunities exist for businesses that can figure out how to keep the pipes flowing.
>
> Yes, it's a hot market. So, expand your vision for a minute. How many bottles of water do you drink a week? How much did you use for a shower? When you flushed a toilet? Wash your car? Cooking? Lattes? And my guess is your city water bill's gone up in recent years.
>
> So ask yourself: What happens in the next 40 years when another three billion people come into the world? Imagine adding 75 million people every year, six million a month, 200,000 every day, all demanding more and more water to drink, to shower, to cook, to everything. All guzzling down the New Gold that's getting ever scarcer.[10]

The social justice people are not the only ones to find this distasteful. A combination of gloating and ignorance is never very pretty.

Meanwhile, if you want to cash in on the new gold or the blue gold or whatever the enthusiasts are calling it, there are many ways of doing it. Below, a look at the world's biggest water companies.

Suez Environnement of France, was spun off from the much larger Suez SA and is primarily an electricity and natural gas supplier. Its revenues in 2011 were slightly better than 12 billion euros, earning a profit of 533 million. Among its many subsidiaries you can count the Barcelona water-treatment and supply company, a number of Chinese water-treatment companies, Fairtec, and Lyonnaise des Eaux, which runs utilities in France and many others in multiple countries. In the United States, Suez owns United Water, the former Hackensack Water Company of New Jersey, and has contracts with half a dozen Canadian municipalities, including Montreal. Suez ran the water system in Buenos Aires for a few years but pulled out when rates were frozen. The company has been actively pushing the notion of public-private partnerships that the cynical interpret as a way for private companies to get their mitts on taxpayers' money. (In at least one case, a private corporation actually filed for status as a municipality under Canadian law so as not to be taxed; it was denied. Jim Southworth, president of Consumer Utilities, expressed his dismay that the Canadian tax code seems, inexplicably to him, to favour public projects.)

Veolia Environnement, also of France, grew out of a company created by Napoleon called Compagnie Générale des Eaux and was recently spun off from the ailing conglomerate Vivendi. Veolia generates most of its revenue in Europe. It was expanding into the United States but has recently backed off; it is still present in the Canadian oil sands. Veolia also operates energy and transportation services, and manages waste in a dozen countries. In general, it now prefers to manage rather than own. Its 2012, revenues were around twenty-nine billion euros, with profits of a little over a billion euros.

RWE AG of Germany, Europe's biggest power generating company, assembled an international water empire in the first decade of this century that included American Water in the United States and Thames Water in the United Kingdom — both now divested, as RWE scurries out of the water business. RWE's expansion was typical of the heady enthusiasm of the past decade: in its 2001 report, the company appropriated the phrase "blue gold," and paid a huge premium for American Water after declaring the United States "the world's most attractive water market," ponying up $7.6 billion and assuming $3 billion in debt. After RWE bailed, American Water was left with more than $1 billion in "legacy costs." Nevertheless, the company still serves fourteen million people in thirty US states and two Canadian provinces. It is well known in the anti-privatization world for its active lobbying against regulation and its creative use of "surcharges" to get around regulated price ceilings. As of 2012, it found a new revenue source to its liking, and began laying water pipes to serve natural gas drillers and oil frackers.

For its part, Thames, now on its own, supplies water to more than a quarter of the British population. It is tightly regulated after a rocky start in the Thatcher years, when it was accused of price gouging and cutting off customers who couldn't pay. Other English water companies are Biwater, Severn Trent, and Anglian Water. In 2014, the British government introduced more competition to the water business in an effort to improve customer service. As of 2017, new water companies will be able to tap into the incumbents' networks for a price agreed upon by the national regulator and sell the service downstream, much as in the market for broadband or mobile phones.[11]

For investors queasy about ethics, a little nose-holding was usually required to invest in many of these corporations, who were seldom scrupulous about their partners. In 1998, to take one example, the British company Thames Water and the French firm Lyonnaise des Eaux had their contracts to manage Jakarta's water supply suspended after their private contracts with cronies of the dictator, Suharto, were revealed.

But direct investment isn't the only way of entering water markets. Standard & Poor, Bloomberg, and Dow Jones have all set up indexes to track the performance of companies involved with water. And dozens of mutual funds have water portfolios.

Some companies trade directly in water rights, in those few jurisdictions where it is possible. WaterBank of New Mexico is one of them, buying, trading, and selling water rights, springs, bottled water, and water utilities, mostly in the southern United States but also in British Columbia. Another is the Summit group, operated by a former CIA analyst, John Dickerson. Summit started in the conventional investment way, by buying into companies that made equipment used in the water business — so-called hydro-commerce companies. But Dickerson soon tired of having a limited number of companies as a target, so he founded a separate hedge fund, Summit Water Development Group, with a very different purpose. It accumulates water rights from willing sellers, mostly in the Colorado basin and along the Murray-Darling system in Australia, not coincidentally places that have recently experienced monumental droughts and among the few places that permit trading in water. The aim was, and is, to resell his accumulated rights to parched towns up and down the basin, a market that looked ever surer every year as municipalities began running out of easily reachable water.[12] Dickerson calls this "wet water," as opposed to the hardware for shifting it around and cleaning it up.

Does all the foregoing constitute the anti-privatizers' much-feared "global water cartel" that will keep prices high for ordinary users? A cartel is, after all, an association of manufacturers or suppliers that agree among themselves to maintain prices at a high level and to restrict or eliminate competitors. In that they differ from monopolies, in which one supplier crowds out all the others, sometimes with government and community blessing. Electricity supply was one such legalized monopoly before competition was introduced in some regions, and some water utilities are run along the same lines.

In most industrialized countries, cartels are illegal — except where they are allowed. Major League Baseball in the United States is an example of a legalized cartel. Canada's supply-management system for agricultural products is another, though it is not *called* a cartel by the governments that regulate it. So in a way, cartels can be relatively benign. The rationale is that it makes for stability of supply and insulates small suppliers. In popular definition, though, there is nothing benign about most cartels, which are correctly identified with, say, the drug lords of Columbia and, more loosely, the Mafia.

In almost all aspects of the supposedly trillion-dollar business of water, there is clearly no cartel. The vast majority of investments in water, and companies involved with water, have little or nothing to do with supplying the actual stuff. And in all water sectors, there is unrelenting competition.

For example, when the Southern Nevada Water Authority, facing critical water shortages, decided in 2008 it must build a third and deeper intake pipe to Lake Mead, it never contemplated building the tunnel itself. The authority put out a call for tenders and got more than a dozen bids, including one from Bechtel. In the end, the contract was awarded to a consortium formed for the purpose, called Vegas Tunnel Constructors, owned by the Milanese construction giant Salini Impregilo and its American subsidiary, the S.A. Healy Company.

A cartel would have prevented competitive bids, in exchange for an uncontested project elsewhere — that's what cartels do. But in this case, there were plenty of competitive bids.

The same is true in other aspects of the water business. Pipeline contractors, desalination plant builders and operators, and others routinely bid against each other. So far, no one is opposing ongoing efforts to tanker water from Alaska to the parched Southwest, but not because it is "cartel policy," but rather because everyone else believes the proponent is going to lose his shirt. In any case, the project continues to languish.

In none of this is there any evidence of price-fixing collusion or restraint of trade by closing out competitors, both hallmarks of true cartels. I am not trying to assert that privatization is entirely benign. It is just that worrying about cartels is a distraction from the three main, if narrower, issues at stake.

The first is how much water corporate interests are allowed to extract from publicly owned reservoirs and aquifers, and at what price. Several examples have raised local and activist ire. The most prominent example in the anti-globalization literature concerns the Coca-Cola plant in the Indian village of Kaladera, in the state of Rajasthan, where groundwater levels dropped from nine to thirty-eight metres in the two decades leading up to 2006, increasing costs to local farmers and indirectly reducing milk yield. That became an easy target: Who could argue for sweet sugar-water over milk? Still, Peter Gleick has pointed out that (a) the water table was dropping before the plant even opened, and (b) even at the height of production, Coca-Cola extracted only somewhere between 2 and 8 percent of available water, but it got blamed for all of the losses.[13] In a similar Indian case, Coca-Cola ponied up $48 million in compensation after being accused, on thin evidence, of damaging community water supplies in the village of Plachimada, in Kerala State. There are also examples in rich-world countries: a bottle-water company was accused of depleting an aquifer near Montreal; and the Campbell Soup Company in Napoleon, Ohio, supplies itself with vast quantities of Maumee River water to make its products, at no cost to itself. The company defends itself by pointing out that other users pay nothing for the water they extract from the river either.[14] We need to consider seriously if this kind of free-water access can continue.

The second of the narrower issues is whether private interests should be permitted to hoard water for later resale at a profit. Compared with the first, this one is conceptually trickier and ethically much more dubious, if rather restricted — in only a few places is such stockpiling permitted. An example is Mesa Water, a corporation owned by T. Boone Pickens, who we have already met. Pickens has been a bit of a gadfly in the resources business for years: a decade ago he was floating the idea of a wind corridor of huge wind turbines stretching from Texas to the Canadian border; he frequently turned up on television to promote the Pickens Plan, though he never seems to have pitched the idea to harder-headed audiences, like investors. In any case, Mesa Water has accumulated eighty thousand hectares of rangeland in Roberts County, Texas, not because Pickens's heart is in the ranching business but because the land comes with associated groundwater rights. (Remember the easement Pickens was obliged to get in order to ship the stuff out of state or to parched Texan cities, like Lubbock, since after it appears on the surface, under Texas law he could no longer own it.)

Another example, from Texas again, is the fate of an aquifer called the Carrizo-Wilcox aquifer in rural central Texas that — or so say the water marketers who buy water rights for resale to cities — holds "many trillions of gallons," enough to sustain Austin and San Antonio for, well, "centuries." Opponents, some of whom actually live above the aquifer, supported by a clutch of environmental groups, say that the bundlers' plans to divert 190 billion litres a year will drain an already depleting resource within decades. This has raised awkward questions. Who is the water for? Who has, or should have, priority? How much will cities drive up the price? Will cities outbid farmers? And who will be compensated if they can no longer afford the water? Some of the water marketers have already offered compensation to landowners to enable them to drill deeper if their water tables drop as a result of over-pumping, but the level of compensation remains an issue. And — a piquant detail this — one of the water resellers, Aqua Water Supply, has filed lawsuits against other such vendors, on the grounds that *their* pumping would impact Aqua's ability to serve *its* customers.[15]

John Dickerson's Summit hedge fund is doing the same thing on a more decentralized scale. Neither Pickens nor Aqua nor Dickerson are buying the actual water — instead, they are buying the right to sell rights to water. In Dickerson's case, as explained above, he has bought historic (i.e., first-in-time) rights to water from creeks, ditches, and canals, previously owned by pioneer families who no longer needed the water, or by successor owners who had bought the same rights earlier and now sought to take their profit. Rights to surface water are more complicated than are rights to groundwater, mostly because if you don't use the rights, or sell the rights, someone else downstream with second-in-time rights will be able to use yours too. On the other hand, the resource will still be there when the time comes to sell, because creeks replenish themselves — or they did before the major drought set in.

You can see why water reselling (hoarding) causes anxiety: Where is the morality of allowing profiteering on a substance that is entirely necessary for life? What happens if a city or town or neighbourhood or even a family is desperate for new supplies because its own are drying up, and the hoarder charges it extravagant prices? The classic free-market argument is that water will flow to those users who value it more highly, and that therefore its use will become economically more efficient. Which is precisely the point: What about poor users who nevertheless need water and can no longer afford it?

Of course, there are solutions to all this, and we will get to them presently. And this hoarding is not as exploitive as it sounds. For one thing, the water hoarders are hardly ever the only source of water and therefore can't really control the price. There are always other sellers elsewhere, and there is always the option of "making" more water through desalination — so the hoarders cannot just charge whatever they please. And on the plus side, these hoarded reservoirs do provide a backup supply of water if needed. In addition, state water regulators still get to rule on pricing. Sellers have to apply for permission to raise prices, and such permission is not always forthcoming. So there are hedges against gouging, and as the drought is prolonged and supplies seem vulnerable, those hedges are getting stronger.

The third issue is at the heart of the matter, for what anti-privatization activists are really talking about is not the water business but what the telecommunications industry calls the "last mile" business: who controls how water gets to ordinary people, at what volume, in what condition, and at what price.

In this the water behemoths seem to be losing.

An instructive example, reported in the *Wall Street Journal* by reporter Mike Esterl, is from the California town of Felton, a community of fewer than five thousand in Santa Cruz County. Felton is on the north side of Monterey Bay, near but not on the coast. It is prosperous enough but not affluent, pleasant enough but not the most picturesque place in the state, though it does have groves of California redwoods that lure tourists. Its municipal water delivery system has been privately owned since the late 1800s, unlike the vast majority of US systems, which are still operated by public utilities. It was part of a larger operator called Citizens Utilities, based in Connecticut.

In 2001, just when the water world's privatization drive was getting seriously under way and bigger players were entering the market, Citizens Utilities, and with it Felton's little system, was bought by a company called California American Water, or Cal-Am, a subsidiary of the New Jersey company American Water. Shortly after that, American Water was itself swallowed by the German electricity supplier RWE, then intent, as we saw earlier, on global acquisitions in the water business. For efficiency's sake, RWE in turn wanted to merge Felton's system with that of Monterey around the bay to the south, which it had picked up in the same acquisitive binge. It wanted to do a lot of other mergers around the country too but soon learned something that due diligence would have told it earlier: that simply to buy American Water required bureaucratic approvals from more than a dozen states, and that further approvals, often from the state but also from local communities, were needed to raise prices. The sale went through — after sixteen months. Price hikes, not so easy.

For Felton, the proposed merger with Monterey was a minor blow to community pride, and if that were the only change, no one would seriously have opposed it. But as soon as they acquired operating control of the company, Cal-Am's new owners applied to the California Public Utilities Commission to raise prices by 74 percent over three years, suggesting that the cost of infrastructure repairs made the hike necessary.

Within weeks, a group that came to call itself FLOW, for Friends of Locally Owned Water, launched an advertising blitz and community-organizing drive, opposing both the rate increase and the merger, and demanding that Santa Cruz County set up a public agency to run the water districts affected — that is, that they "re-municipalize" water. Among the complaints was that American Water had centralized its operations, and to complain about a broken water main or some other minor problem meant phoning a call centre in Illinois, two time zones away. The company rather plaintively said that the call centre idea was to boost service, not restrict it, but by that time, no one was listening. After more than a year of mulling it over, the state utilities commission granted a 44 percent rate increase, still far too much for FLOW.

With opposition mounting, Cal-Am's response became a textbook case of how to do things the wrong way. A prickly press release was issued saying the company was not for sale to anyone at any price, including to the county or the town. A county supervisor up for re-election, who was thought to be in favour of FLOW, was told his election would be "torpedoed" (his word) if he supported a public takeover (the company denies any threat, but nevertheless sent off mass mailings urging people to avoid voting for, well, people like *that*). Thousands of dollars were paid to a local property association that opposed the notion of re-municipalization, arguing that homeowners would face massive price increases if the public option went through.

In July 2005, the FLOW-sponsored Measure W was overwhelmingly approved in a referendum. The measure approved the issuance of $11 million in new bonds for the explicit purpose of buying the Felton water system. It even approved a tax hike to do so, which Cal-Am had assumed would scuttle the bid. When it didn't, the company once again announced that nothing was for sale. This was a little disingenuous, since in November of the same year, RWE bailed out of American Water altogether and basically out of the US water market entirely, citing considerable political resistance to privatization in the water sector as a reason for going home to Europe. There had been reactions similar to Felton's in a slew of other American communities, among them Monterey itself, Urbana, Illinois (which was facing boil-water advisories as water quality deteriorated); Chattanooga, Tennessee; Montara, California; Stockton, California; and Lexington, Kentucky.

The Felton saga wasn't quite finished. The town's water system was still privately owned, though once again by an American company, not a German one. A new purchase offer was spurned, so the community turned to the notion of eminent domain to precipitate a buyout. It took two more years of litigation and a hotly contested valuation for the property before Cal-Am finally conceded. There have not been massive price hikes since, but water rates have gone up.

RWE had faced problems elsewhere, not just in the United States. Its other major water property, Thames Water of Britain, was under fire for raising rates without fixing infrastructure problems. RWE, in a kind of *Jeez, who needs this?* reaction, put that up for sale too. The company also made a hurried exit from Shanghai after its guaranteed profit margin was arbitrarily reduced. All over the world — RWE water divisions operated in forty countries, an empire than took years and billions of dollars to put together — the apparatus was being dismantled. As the *Wall Street Journal*'s Mike Esterl put it, RWE found that "water turned out to be less like electricity than RWE hoped. It's heavy and hard to transport, making it difficult to build economies of scale. Regulation is never predictable."[16] *That* was something of an understatement.

It is not just RWE, either. Suez, most famous for its debacle in Atlanta (a twenty-year contract ended after four years, after residents of even affluent neighbourhoods found brown sludge coming out of their taps), has been backing off bidding for contracts in the developing world, and Veolia has done much the same.

The real reason is only indirectly political. Water, and sanitation with it, turns out to be an infrastructure sector with a high ratio of capital to revenues, with a return-on-investment period that is long, in an environment that is politically sensitive and easily inflamed. Privatization in these circumstances, then, turns out to be a high-risk, low-return investment. As John Briscoe told me, repeating a point he had made many times, "In a typical concession contract there is negative cash flow for about ten years, and then positive flows for the last twenty years. But if a contract is annulled after ten years, the investor loses everything. The credibility [and stability] of a government is thus critical.... The number of concessions in telecoms which have been annulled is just 3 percent; in electricity, 8 percent; and in water, 33 percent."

All kind of investors in water, many of them vainly hoping Big Water will turn out to have Big Profits, have found out — to their cost — that water is not, after all, a commodity like oil. Instead, the asset's market value depends more on political will than on demand, and in this case, the political will is uniquely vulnerable to citizen persuasion.

There's another aspect to the relationship of the corporate world to water that has gone largely unnoticed, and it is that growing numbers of corporate executives and their companies are beginning to understand that water — or the lack of it — is about to become a major corporate problem. As the *Economist* reported in 2014, "An increasing number of bosses say water is or will soon become a constraint on their firm's growth. They are right to worry, but most firms are not doing much about the problem."[17] At risk are not just companies whose products are largely made of water (beer, for example) but also companies whose products depend wholly on it (food producers), as well as those doing business in water-stressed countries like China.

Privatization's Pros and Cons

In some places, privatization has worked. In others, not so much. It is useful to look at a few examples.
Originally, I had São Paulo chalked up in the win column for privatization. Clearly, losses can be converted to wins. And vice versa.

Manila, the Philippines

You can make the case that the privatization of municipal water systems, at least as envisaged by the World Bank and other international lending institutions, had its origins in the Philippines capital. The water system in the city was a mess. A large percentage of the population had no piped water at all, and nearly half of what water did flow through the system was lost to theft and leakages. Illegal siphoning caused sporadic outbreaks of cholera through contamination of the supply. It is also fair to say that this was fairly typical of public services in the country — the electricity grid was, if anything, worse. The two decades of misrule by Ferdinand Marcos, starting in 1965, were characterized by repeated efforts to upgrade infrastructure, in every case sabotaged by rampant corruption. Four times between 1960 and 1990, the World Bank put up money to fix the system, and four times it failed. The first effort was to try to reduce water losses from 45 to 35 percent, but by 1990 the losses were 65 percent, not exactly a stellar record: after two decades of investment, everything had got worse. The bank had had enough. As John Briscoe put it, "Fundamentally, we realized that without a change in incentives — some very logical, sensible things — this was not working."

This belated recognition coincided with the fall of Marcos and the election of Fidel Ramos as president. In his early years, Ramos used energy concessions to private contractors to fix the electricity grid, with considerable success. He and the bank believed the same might work for water. His staff rewrote the legal and regulatory framework and, with World Bank support, let two bids for concessions in Manila. At first, it seemed a failure. The concessionaire in east Manila went broke almost immediately, without fixing anything, wrecked by a rapid devaluation of the Philippine peso. A new consortium took over, 65 percent owned by a local conglomerate, Ayala, partnering with United Water and Bechtel. In fairly short order, over two million residents were hooked up to the grid and water losses were down to 30 percent, better than many systems elsewhere (London's, for example, was 42 percent). It is true that the system is far from achieving 100 percent coverage. It is true that the rates have gone up, sometimes steeply. It is also true that the poor are charged a tariff for water, and not a trivial one. On the other hand, it is true that many of the poor now have piped water for the first time, and true too that the rates they pay are far lower than they were paying to private water vendors before. It is true, finally, that the system is self-sustaining.

Count this one a success.

Dar es Salaam, Tanzania

No success here.

The city government was pushed into privatization, or jumped into it, depending on who you believe. Both are probably true. On the jump side, media reports at the time suggest that the World Bank's offer to put up more than $160 million to privatize the supply system was met with relief. It was hardly a secret that the water system was a mess, and even less of a secret that the national government had no money to fix it. Opponents suggest, by contrast, that Tanzania was bullied into privatizing through private and World Bank pressure. This, also, was probably true — the World Bank had no reason to throw any more money at a corrupt and inefficient bureaucracy, and had no reason to suspect that private operators would be worse.

In August 2003, the project was put out to tender. The winning bid was a consortium whose main partner was Biwater of Britain. Gauff Tanzania, a subsidiary of a German consultancy, was a junior partner, along with a Tanzanian company that makes trailers, the Superdoll Trailer Company, just going along for the ride. The bidding process saw many of the classic mistakes of failed privatization contracts elsewhere. There was no prior public consultation and, indeed, the contract was not disclosed even to the national parliament until after it was signed. Instead, costs were kept below full-recovery levels by no-interest loans and a few sweetheart construction contracts on the side. A level of public deceit was therefore built in from the start. There was only one bid, quickly accepted.

As the writer Martin Pigeon explains, problems began almost immediately. The consortium called City Water Services (CWS) contributed only half the capital it was supposed to, and within five months had stopped paying its monthly lease fee to the municipal authority that was supposed to benefit, the Dar es Salaam Water and Sewerage Authority (DAWASA), the given reason that what money had been paid had somehow been "disappeared." What little efficiency had existed seemed to collapse entirely — new customers weren't hooked up but got bills anyway, existing customers continued to get water but never got a bill. Promised construction projects never began. CWS provided no financial reports to its overseer. Just two years after the ten-year contract was signed, the Tanzanian government shut it down. Biwater-Gauff sued, looking for $20 million in damages. Both the UN Commission on International Trade Law and the International Center for Settlement of Investment Disputes found that Tanzania was technically guilty of illegally abrogating the contract but imposed no financial penalty, so egregious was CWS's "fulfillment" of its conditions.

So DAWASA, now renamed DAWASCO (Dar es Salaam Water and Sewerage Corporation) regained control in what was a "parastatal company owned and financed by the national government." The system remained a mess. Three-quarters of the supplied water couldn't be tracked or accounted for. Slightly more than 60 percent of city residents used piped water, but only 8 percent had taps in their homes — the rest used communal block taps. Connected users got water about eight hours a day. Leaks and illegal connections, along with an increasingly limited (and polluted) supply from the Ruvu River, meant that the system could supply only about a quarter of the demand.[18]

Matters have improved somewhat. Connections remain low but have gone up 12 percent. More users are being billed, and tariffs, while they have gone up 30 percent, remain low. But capital expenditures are not being made, necessary construction goes unbuilt, staff morale is abysmal, expertise is lacking, layoffs continue, and corruption is unabated. (Most of the illegal connections are actually performed by DAWASCO employees, and overbilling of those that can pay is endemic.)

Most residents continue to rely on private water sellers, who charge high prices. This is de facto privatization of another sort.

Consider this one a failure on all fronts.

Atlanta, Georgia
This was going to be the poster child for privatization. United Water, the Suez subsidiary that won the twenty-year, $428 million contract to operate Atlanta's water and sewage system, boasted that "Atlanta will be for us a reference worldwide, a kind of showcase." Suez was

going to cut costs to everyone, improve revenues, and end the quality problems that had been plaguing Atlanta for decades. It would be a model for the hundreds of similar contracts the water companies confidently expected would be signed as a result.

Atlanta certainly wasn't bullied into offering this contract, not by the World Bank or anyone else. It wanted it because the capital's water and sewer systems were in disrepair and steadily declining. There were so many outraged communities downstream of Atlanta's sewer outfalls that the city was obliged to pay penalties of nearly $40 million in recompense. The resource was so mismanaged that, in 2007, Atlanta itself came within two months of running out of water entirely, and the governor, risibly, was reduced to wondering whether it was constitutional to pray for rain. A multinational with deep experience and plenty of capital would surely do better.

The contract was signed in January 1999. It was terminated with cause in January 2003. In retrospect, the company had over-promised: it guaranteed to radically reduce operating costs and to extend environmental protections, promises that, it turned out, it had no way of keeping. Sure, it could reduce operating costs by paring staffing to the bone, but that simply caused an escalating series of quality problems. A major water leak went unrepaired for two years and was fixed only when the state threatened to hire a contractor to fix the problem and charge United Water the cost. In May 2002, the state Environmental Protection Agency was forced to issue boil-water advisories, even to the affluent North Buckhead neighbourhood, where the tap water turned rusty and contained unsightly unspecified brown debris.

In consequence, Atlanta's water has been "re-municipalized." The anti-privatization NGO Water-Allies, an offshoot of Food and Water Watch, has claimed at least some of the credit. But the short explanation is that the company screwed up. So much for the vaunted efficiency and expertise of corporate entities.

No question about this one. Not just a failure. A rout.

Australia

The world's driest continent is also the world's most freewheeling water market. It was also subjected to one of the worst droughts faced by an otherwise productive agricultural sector, the Millennium Drought of 1997 to 2009.

The country started its liberalization drive, which included privatizing formerly state-run enterprises and services, in the 1980s, well before the drought began. This was extended in the early 1990s, again before the drought, and water rights became a commodity to be freely traded on an open market. As in the American Southwest, water was "delinked" from the land it flowed through. Anybody could hold a water right, not just other riparians, with "anybody" defined as persons, farms, corporations, or municipalities. By extension, rights holders could sell their allocations to anyone they pleased, even foreign investors, at whatever price was agreed to by both parties — just like in any other market. There were hiccups along the way: it wasn't always clear, for example, who owned what rights before the change; entitlements were often poorly documented or not documented at all; there were sometimes competing claims to the same water; and, at least by one measure, there were more entitlements on, say, the Murray River than there was water in the first place. It was also tricky to measure actual water flows in rural areas, from which allocations could be made. At the height of the drought, this was sometimes complicated by what amounted to water theft or water fraud — clandestine hoses and camouflaged illegal dams, pumps running where they shouldn't. The Australian Competition and Consumer Commission was given the power to develop and enforce the market rules, in effect becoming water cops in addition to regulating the market itself.

In the end, rights were established with a minimum of rancour, based on prior use and seniority. Thereafter, trades were made largely through brokers in the city of Adelaide, the nearest city to the mouth of the Murray-Darling, the continent's greatest river system. Prices were established by buyer and seller but affected also by evaporation rates, seepages, and other natural forces. Water companies such as John Dickerson's Summit and the Australian Causeway Water Fund would sometimes accumulate their own water rights, or find buyers for them elsewhere. In the early years, Summit bought ten thousand megalitres that it planned to stockpile and perhaps rent rights back to whoever would pay.[19] An Australian investment bank, Macquarie, bought water rights for its investment in almond groves. In 2007, as much as 40 percent of the available water was on the market.

All this caused great alarm among opponents of privatization, which feared, with some justice, that small farmers would be squeezed out and big agribusiness allowed to muscle in. They had a point. But it should be remembered that the water itself remains in the river — a Hong Kong investor may buy the rights but would have to find a willing buyer among those who could actually use the water. That investor couldn't take the water elsewhere. And the investor did face competition from other sellers, with the result that prices couldn't escalate at whim. Sure, a hotel in Adelaide could pay more for an allocation than a farmer, but then, that's what markets are for — the seller could make some money getting rid of something no longer needed. At the height of the drought, many beef farmers, for example, shut off their pumps and sold to someone downstream, often wine farmers. In this way, they would no longer need to irrigate fields to grow crops for their cattle; instead, the sale would give them the money to buy feed elsewhere. Some farmers went bankrupt, but on the whole, economic damage was minimal.

At the height of the drought, available water dropped in the Murray-Darling basin by a remarkable 70 percent, and yet it had minimal impact on agriculture. There were shifts in crops — as water availability went down, prices went up, so the rice farmer (who used lots of water for a low-value crop) would sell to a grape grower, whose product is more valuable; the rice grower got more money than he or she would have growing rice — so there was no one left needing compensation, and nothing for government to do. Two years after the drought ended, agricultural production once again neared pre-drought levels. As a piece in the journal *Science* argued, producers were buffered by "(i) well-developed water markets that allowed water trade to farmers in the greatest need; (ii) modernization of irrigation infrastructure that increased the efficiency of water delivery; and (iii) establishment of clear water entitlements for the environment that protected critical refuge habitats and populations as water availability declined."[20] In effect, water trading encouraged conservation, because water became more valuable, exactly the opposite of what anti-privatizers feared.

Perhaps it was successful because there were reasonable checks and balances on the freewheeling water market. The actual water service (the plumbing) continued to be operated by public utilities, and as the *Science* piece noted, the legislation that set up the system of tradable usufructuary rights reserved a considerable quantity of water for ecosystem services — that is, for the river itself and its "environmental flow," whose preservation remained a high priority. At the same time, farmers were essentially told they could no longer expect drought and debt relief from the government, forcing them into conservation. Even so, it remains true that agriculture, which amounts to less than 3 percent of the Australian economy, uses two-thirds of the available water supply.[21]

Several other developments helped Australia get through the drought. One was that municipal water use was reduced by nearly 50 percent largely through simple water-use restrictions on activities like lawn watering and car washing that were generally understood and well received by the public, since usage levels have not rebounded after the drought. The other is the rapid construction and deployment of large desalination plants. We will come back to desal in Part Five, on solutions, but it is worth noting here that the city of Perth was, within two years, able to envisage and build a plant that pushes more than 90,000 litres of purified sea water a minute (130 megalitres a day) into the mains. Other plants in New South Wales and Queensland are almost double the size.

The Australian story shows us that, in a world facing increased and prolonged droughts, conservation should become the highest ethic. It also shows us that fully tradable water rights got the country through the drought with a minimum of damage.

Paris

The French capital is also the headquarters of the world's two largest private water companies, Suez and Veolia Environnement, and their ouster from running Paris's water and sewage is a body blow to their credibility. In this case, though, the abrogation of the contracts (Veolia had the right bank, Suez the left) was driven neither by anti-privatization NGOs and activists, nor by public dissatisfaction, but by the peculiar, inturned, arcane nature of French politics. And it was complicated, in turn, by the peculiar, inturned, arcane character of French trade unions.

Nothing was simple from the beginning. For centuries, water and sewage services had been owned and operated by the prefecture of Paris; the city's sewers were, rightly, greatly admired by engineers and city managers everywhere and are still a tourist attraction — guided tours are offered to the fascinated and the non-fastidious. In 1985, however, in the same flush of enthusiasm for neo-liberalism on offer elsewhere, the water distribution network was privatized; two years later, so was the water production system, or at least partially — the city owned 75 percent of that, the two multinationals got 12 percent each, and an investment bank the small portion left. Sewage remained entirely in public hands, so to speak.

Generally, this new arrangement did well. Leaks, which were already low by big city standards (around 22 percent), were reduced to 13 percent and then to 3.5 percent, an extraordinary figure. Still, water bills went up at a rate far exceeding inflation, and when the private companies, jealous of each other and zealous in guarding proprietary information, neglected to provide public accounts, political suspicion grew.

In 2001, Paris elected a socialist mayor with the declared aim of taking the water system back in-house. Neither Veolia or Suez took this very seriously — they made a few concessions here and there, offered to lower their profit margin to around 4 percent and to invest more in infrastructure renewal. In this insouciance, they were in error: Mayor Bertrand Delanoë repeated his threat in his second successful election campaign in 2007. In July 2008, the deed was done.

The transition wasn't easy. Between the private companies and the city, no fewer than fifteen unions had to be placated, and contracts renegotiated. Senior executive fled, and the city didn't have the money to hire them back, so a management vacuum followed. The city's finance department, which had not taken the mayor's promises seriously either, had failed to make any preparations. And under the complicated procurement system of public Paris, all decisions had to be pre-cleared by a commission and then reargued by the board (which contained NGOs and union reps, and well as politicians), resulting in endless delays and frustration. All management hiring was, and is, done directly by city council.

And yet consumers hardly noticed the difference. Excellent water supplies continued as before. Paris sewers did their job as they always had. Even better, the "re-municipalization" saved money — it turned out that the city, with all its cumbersome structures, was a better manager than the multinationals. In January 2011, the water utility, now called Eau de Paris, actually lowered water rates by 8 percent, a humiliation for the private companies that had insisted higher rates, not lower ones, were needed. Even more annoying and embarrassing for Suez and Veolia, the utility has issued ringing declarations against bottled water, has subsidized the poorest forty-four thousand households in the city, and has consistently refused to cut off water to scofflaws, the poor, and those in squats, all this on reduced rates.[22]

It's easy enough to find many other places where privatization has worked for consumers as well as for the companies themselves — Chile is an example, as is Phnom Penh in Cambodia, Casablanca in Morocco, Karnataka State in India, Bechtel's operations in Tamil Nadu, Suez's in Delhi, and Veolia's in a dozen cities. Gabon is a good example of a projected countrywide success with Bechtel the contractor. Senegal is another. In cases like that of Bogor, Indonesia, prices were raised substantially in the first year of privatization, but the result was that the grid was extended to many thousands of the poor who had never been reached before, and gripes were minimal.

It is also easy enough to find failures. Suez pulled out of a contract in the Turkish town of Antalya after protests over 130 percent price increases; Guyana cancelled a contract with Severn Trent for non-performance, as did Nepal; Suez has abandoned its projects in Argentina; the city of Quito in Ecuador cancelled a privatization contract when consultants pointed out that it would guarantee a profit to companies that would invest only $7 million compared to the city's $226 million; Vietnam has terminated its sewage contract with Suez. Many failed for the same reason that Quito identified: poorly written contracts. In other cases, movements similar to the one in Felton, California, derailed privatization efforts, sometimes for good reasons, sometimes not. Portland, Oregon, is a case in point: early in 2015, there was an uproar as the city council assigned a private company, CH2M Hill, to dismantle hundred-year-old reservoirs (elegant enough to be on the National Register of Historic Places) and build new bunker-like ones, covered instead of open. Apart from a great public affection for the older reservoirs, the real issue was one of cronyism. One of CH2M Hills's vice-presidents was a former chief engineer with the "— and one of the prime consultants to an Environmental Protection Agency report that argued for the new reservoirs in the first place. The city's water managers plead that safety is their prime concern, but their own record is hardly pristine: they have been penalized by state authorities thirteen times since 1998 for water-quality violations, paying fines of around $750,000. At the time of writing, the outcome was still uncertain.[23]

Further, it is *also* easy enough to find cases where privatization is working well, at least in the sense of delivering on its promises, but where governments undercut profitability and the viability of the contracts by unrealistic and politically driven pricing. Jakarta is an example on this side of the ledger, though Uruguay is perhaps the best example of privatization succumbing to politics. After years of fairly uncontentious private management, a national referendum declared that water was a basic human right, to be delivered at no profit by public entities — a decision that was later incorporated in the country's constitution. Politics were served and services stayed pretty much the same.

Water Services to the World's Poor

One of the charges against privatization is that "untold millions in the Global South" have been cut off from access to drinking water because they either couldn't afford new rates or simply didn't pay their bills. Dozens of stories are still circulating that private water companies boosted prices beyond the reach of the poor, and even cut off some of the truly indigent altogether. Where are these untold millions to be found?

Most of the stories I tracked turned out to be false. But not all.

Untold thousands, maybe.

Some of the untold thousands, and then only for a brief period, were to be found in South Africa, one of the few countries other than Uruguay and Bolivia to have declared water a basic human right. As a backdrop, it should be remembered that when the African National Congress (ANC) inherited the country from the last apartheid government, twelve million of a total population of thirty-eight million had no access to clean water. By 2002, that number was down to seven million of a total population of forty-two million. The shantytowns outside the major cities were ghastly still, but there was piped water to public standpipes and basic latrines for the first time. It may not have looked like much to outsiders, but it *was* progress. The government has achieved this remarkable result through a combination of political determination and a willingness to experiment, devising policies that initially met opposition from both the development banks and NGO lobby groups. It was done by judiciously involving private companies in water management. David McDonald, co-director of the Municipal Services Project, which has been monitoring the privatization of water services in South Africa, says privatization "is most often seen as part necessity, part political choice because of fiscal constraints," reducing public borrowing, taxes, and outlays.[24]

Each household in the country, regardless of income, is supposed to receive a basic allowance of six thousand litres per month — not very much, but enough for thrifty use by a smallish family and a lot better than they ever got before. However, in some municipalities, such as the Empangeni region of KwaZulu-Natal Province, prepaid meter cards were introduced in 1998, and those who couldn't afford them were forced to dip their buckets into local rivers, and drink the polluted river water. The result was a sudden spike in cholera cases, first to ten thousand and then to five times that number. The same thing happened in Phiri township, a part of Soweto. But there, a group of residents in a community long accustomed to collective political action, rebelled — and then sued. The courts have since forbidden water shutoffs to the indigent, and the meters have been redesigned to allow for the free allocation.

In a few South African regions, the private companies anticipated all this by the simple (but apparently revolutionary) expedient of actually consulting with the people they were supposed to serve. In the 350,000-strong community of Nelspruit, in Mpumalanga (Eastern Transvaal), a subsidiary of the British firm Biwater, encountered thousands of "unregistered connections," made precisely to get around this metering issue. Instead of cutting people off, the company sent rapporteurs into the townships to explain how the system worked, to discuss and encourage the regular payment of bills, and to assure people that they would get a basic allocation regardless. Dozens of local offices were set up where residents could air complaints, report leaks, and generally make contact (the company also undertook street theatre and sports sponsorships to further connect with the population). It worked: leaks were repaired and thirty-three thousand new customers added to the grid. Hardly anyone complained.

More unserved citizens are to be found in Puerto Rico, where the debt-ridden water authority began cutting off supplies to non-payers in the summer of 2014. Among the main scofflaws was the municipality of Arecibo, which had been refusing to pay its $1.5 million water bill. In August, to vociferous protests, the utility began cutting water to non-payers in public housing too. Residents there are charged $19.76 a month, regardless of how much water they use — use is not metered. The utility wanted to up that fee by another $20, mostly to fix what it plaintively calls "deficiencies in the system," a politically improbable idea.[25]

But this can't be laid at the door of privatization. Puerto Rico's utility is publicly owned.

The most notorious case of shutoffs was still happening in 2015 in the bankrupt city of Detroit. Shutoffs started as early as 2002, which was when Marian Kramer, a social activist and head of the National Welfare Rights Union, discovered that the Detroit Water and Sewerage Department had shut off running water to more than forty thousand delinquent customers without offering help or giving them any chance of appeal. All of which led to the bizarre sight, in a first-world country, of hoses snaking from one house to another as neighbours helped each other out, and of women staggering about balancing large buckets of water. The noisy public campaign led by Kramer persuaded Detroit's city council to end the shutoffs and to introduce a Water Affordability Plan to help those who really couldn't pay. And there it rested until 2014, when nearly half the utility's customers were at least sixty days past due — not just people but schools and businesses too. In March, the city again started sending out shutoff notices. Darryl Latimer, the utility's deputy director, told the *New York Times* that "the threat of having service cut will resonate with customers. It's really more a pattern of behavior — 'If you don't come to cut me off, I'm not going to pay until you force me into paying.'"[26] Another outcry ensued, resulting in the odd spectacle of Canadian residents across the Detroit River sending relief tankers across the border to help out, an unedifying sight. The shutoffs stopped, only to resume in August. The UN weighed in, proclaiming that shutting off water to those who can't afford it "constitutes a violation of the human right to water," and arguing that the shutoffs discriminated against minorities. As it turned out, over half the shutoffs were reversed within a day as delinquent customers indeed found the money to pay their bills.

The Detroit shutoffs have attracted a good deal of activist spleen, much of it justified, some of it not. An anti-privatization screed by Ellen Dannin on the website Truthout put it this way: "Americans used to take water for granted, but the water shutoff in Detroit has taught us all-important lessons. We now know that the private sector is willing to be ruthless in denying access to the most basic needs of living beings, and we also know that even those who have the least resources can also have power — if they are organized." But wait a minute. This can't be laid at the door of privatization either. The Detroit Water and Sewerage Department is a public utility, not a private one, though this hasn't stopped anti-privatization activists from having their suspicions. Martin Lukacs, writing in the *Guardian*, suggested that the city's emergency manager, Kevyn Orr, supported shutoffs to make the utility more attractive to potential private buyers. Orr declined to comment, fuelling the suspicions further.

In 2014, the UN, in a little-read report, noted that "courts in India, Brazil and South Africa have reversed decisions to disconnect water when customers don't pay."

Water as a Human Right

Partly as a reaction to, and rejection of, privatization, several countries have adopted the notion that water (or, in some slightly watered-down definitions, *access* to water) should be a basic human right.

The social justice people are pushing this beguiling notion. It seems so obvious, doesn't it? Everyone knows that, without water, you die. What could be more basic?

In 2004, Uruguay became the first country in the world to declare water a basic right. South Africa and Bolivia followed. By 2009, this was such a non-contentious issue that the multinational PepsiCo, driven by shareholder resolutions, publicly committed itself to respecting the human right to water throughout its global operations.[27] A year later, in 2010, Bolivia took a draft resolution to the UN on the human right to water and sanitation. After some hemming and hawing, a resolution was passed supported by 122 countries that declared that "the General Assembly recognizes the right to safe and clean drinking water and sanitation as a human right that is essential for the full enjoyment of life and all human rights," and it suggested that it was the moral responsibility of rich countries to assist poor countries in getting there. Even the United States signed on.

Does this help getting water to the 1.4 billion people in the world without safe access, or getting sanitation to the more than 2.5 billion without it? Does declaring water a human right persuade the governments of these countries to drop everything and focus on getting water to everyone? I wouldn't go as far as Harvard's John Briscoe, who has suggested that "the greatest benefit of declaring water to be a human right is for those people who make this declaration, who think they have done something noble."[28] I think the declaration is probably more manipulative than that: it is part of the bully-pulpit notion that if you declare something important long enough and often enough, it will actually become so, and that will impel people with money to part with some of it for the cause — International Rivers has been using this technique effectively for some time, as we shall see in Chapter 6.

Declarations are all very well, but though the poor have a right in law, they are generally at the very end of the supply lines. The pipes often run dry before the water gets to them, forcing them to depend on private water vendors (not corporations, but guys with trucks) for their supply. It is hardly a secret that too many public utilities in the developing world favour the rich and not the poor; that was true in Cochabamba — the poor are often not hooked up to the grid at all, so the rich get water at 25¢ a cubic metre while the poor buy from trucks at $2 or more a cubic metre.

It is a misconception that World Bank loans favour the private mega-corporations. Over 90 percent of World Bank loans for water infrastructure development worldwide have been to public entities, not private corporations. Contrary to the legend that it uncritically endorses private capital wherever it can, the World Bank has a long record of pouring good money after bad in the water business, making loans to far too many public utilities that operated more like kleptoc-racies than public services. Indeed, the Fourth World Water Forum harshly criticized multinationals not for profiteering at the expense of poor countries but for *pulling out* of non-lucrative contracts, assert-ing that their withdrawal doomed millions to long waits for service. This squares poorly with the counter-narrative that multinationals are cutting off "untold millions" from their precious water, and that privatization has been a catastrophic failure. Well, it has been a failure, but not for the reasons feared.

The Pros and Cons of Privatization

If governments won't, or more likely can't, get water to the poor, who can and will?

Privatization was never altruistic, to be sure — no one is that naive — but in many ways it really was (and is) an attempt to bring expertise and capital where they had never existed before. Whether it works or not depends to a surprising degree on the local governments in place. If they are not serious about making reform work, no reforms will work, whether the systems are private, public, or some form of hybrid. If they *are* serious, almost any system can be made to work.

What Cochabamba proves is that letting multinationals bent on profit loose in a region that is poor, on a contract that is poorly drafted and poorly regulated, is inevitably going to result in conflict.

So how can it be made to work better?

In areas with less than sterling governments, isolating regulation from politicians is a good start, if a difficult one, particularly when politicians line their pockets as well as pad their voters' lists, as the mayor of Cochabamba did. Openness, or transparency, is essential: let consumers know ahead of time what's coming, and don't spring it on them after a deal has been signed, as Tanzania did. Don't insist on everything happening at once — if there are alternate suppliers to hand, let them operate, at least for a while, then integrate them rather than eliminating them. Bechtel in Cochabamba could have done this but screwed it up instead. Prices should rise slowly on a schedule that everyone understands ahead of time, until utility income reaches parity with expenses. There should be support for the poor, a basic allocation at minimal rates — not free, but cheap. Pay attention to training and human resources; good management is not a luxury.[29] Above all, privatization contracts should avoid imposing on the community a moral hazard in which profits are privatized and costs socialized. If a company screws up, it should be obliged to fix it, at its expense. That doesn't sound so hard, does it? But it seldom happens.

Empower regulators to make real decisions. An instructive example is New York State, where regulatory oversight has been keeping a close eye on the private contractors' performance, and is not shy of issuing edicts where they seem warranted. For example, in November 2014, the state's Public Service Commission (PSC) ordered the company supplying water to Rockland and Orange Counties in New York to drop its plans for a desalination plant on the Hudson River that the company maintained was needed to increase supply. The PSC's chair, Audrey Zibelman, "urged" the company, United Water New York, to instead explore cuts in consumption and other conservation methods, and to locate other water sources. The conservation NGOs Riverkeeper and Scenic Hudson applauded the state's decision, pointing out that consumption was already declining because of more efficient appliances and toilets, and that the plant was supposed to be built just a few kilometres downstream of the Indian Point nuclear plant, a less than felicitous location.[30]

Writing proper contracts is one essential for successful privatization. Contracts often contain red-flag clauses that almost guarantee a concession is unlikely to work. One such is the language that existed, for example, in the Lake Ontario city of Hamilton, Ontario, in which the public agency was made entirely responsible for breakdowns in the water system without having any participation in revenues. Under the original contract signed by the city with the contractor, Philips Utilities Management Corporation (PUMC), the city incurred all capital expenditures and major maintenance costs — which soon included cleaning up a sewage spill that affected residential areas and polluted the harbour.

Another red flag is the kind of language now so common in international "free trade" agreements, in which private companies have the right to reimbursement if the public partner does anything, whether by commission or omission, that may deprive the concessionaire of future or even potential revenues. An example from outside the water world is advanced by Ellen Dannin, in a piece on the website Truthout. In this case, a private contractor hired to run a toll road in Virginia demanded compensation whenever car pools (which didn't have to pay tolls) exceeded 25 percent of traffic.[31]

Privatization or not, water needs to be priced properly. As South Africa and other places have amply proved, water can be priced progressively, like income tax — the higher the consumption, the higher the rate. But too often it is the reverse: in California's Central Valley, farmers for many years received volume discounts, instead of paying a premium, though they don't anymore.

On the Côte d'Azur in France, hardly a hotbed of Bolshevism, everyone pays something for water, though not very much. You pay only a little for what is known as "the first glass" — enough water for essential services. If you use more — you like a hot tub, for example — you get to pay a steeper price. Then, if you want to fill one of those infinity swimming pools so beloved of upscale shelter magazines, the sky's the limit.

If the water is priced properly, it doesn't matter what kind of entity delivers the water — public, private, or some hybrid.

In the end, it is hard to see in water privatization the bogeyman so energetically put forward by its opponents. It can be made to work — or not. Private companies are not anti-people, they are just pro something else. Where privatization is introduced, though, it should be with lifeline rates for the poor, strong protection of ecosystem services, and strong regulatory oversight by governments.

My own bias is that governments should provide all basic services: electricity, water, highways and transportation infrastructure, and health services, as well as newer utilities such as high-speed Internet access, and they should generate the tax revenues necessary to pay for them.

It is the opposite bias that declares private capital more creative and capable than government agencies. Sometimes it is, of course, but it's certainly not a given, as some of the examples provided above have shown. As I wrote in a recent book, "You can trot out example after example of corporate executives looting, and then destroying, the corporations they were hired to run, but it doesn't seem to matter. You can show evidence of widespread flim-flammery and corporate malfeasance, of bloated middle management and imperial CEOs, of boards running up unmanageable debt and then crying to the public for rescue, but none of it makes any impression. You can show how many corporations have abandoned the quaint notions of service and quality for numbers-fiddling, hidden accounting systems, speculation, outsourcing, mass firings, and chicanery of the most venal sort, and people will nod their heads and look away."[32] Okay, a little over the top, but someone needs to defend the public sector. Company executives cannot be trusted to police themselves, or act in the public good unless they are made to.

In the rich world, municipalities and regions have resorted to private capital because they are too politically timid to raise the necessary taxes to pay for the necessary fixes. In those cases, with good contracts and decent oversight, privatization can be made to work well. In the poor world, private capital is often the only way to get the infrastructure built in the first place. In those cases, with good contracts and decent oversight, privatization will benefit everyone, even the poor.

6
Are Dams Really That Bad?

Why do so many people so suddenly hate dams? One answer may have to do with the worldview of the people confronting dams. If you are used to freely running rivers, even if they're just burbling brooks, and if migrating salmon or trout are part of your identification with nature, you will be predisposed to see dams as intrusive, as barricades, as obstacles. If, on the other hand, your worldview is shaped by the fact that rivers where you live run only a few months or a few weeks a year and are otherwise dry, you are more likely to regard large dams as insurance, as saving for the future, as risk mitigators, as artifacts that will help keep you alive in bad times. Some of the dissonance, surely, is caused by people from group one extrapolating their experience, and their preconceptions, to group two.

There are good reasons not to like dams, but the opposition expresses a special antipathy not really explainable by fact. As John McPhee put it in *The New Yorker*, "There is a special animus that environmental activists hold for dams. To them, there is something disproportionately and metaphysically sinister about dams. Conservationists who can hold themselves in reasonable check before new oil spills and fresh megalopolises mysteriously go insane at even the thought of a dam."[1]

But the dam story is more interesting than that. It is not just that, in my view, anti-dam people have caused more harm than good by their adamancy (or at least prevented much good from being done) but that they succeeded over the last few decades in bullying the World

Bank into submission. Who would have thought? The bogeyman of the anti-globalization folk meekly submitting to activists with a better sense of politics than they had.

That's changing now. The spell seems to be wearing off. Whether this is a good thing remains to be seen.

In 1900, there were no dams in the world higher than fifteen metres. By 1950 there were 5,270, two of them in China. Thirty years later, there were 36,562, of which no fewer than 18,820 were in China. Best guess is that by 2010 the number had reached 45,000; it's a little uncertain because the metric often varies — sometimes it is said there are so many dams higher than fifteen metres, or higher than five metres, or taller than four storeys. There are said to be in the United States alone 60,000 dams, or 85,000 — the numbers vary, and certainly many of these dams are tiny. Still, our species has built a very large dam about every hour for more than forty years, and has dammed more than half the world's rivers, whose watersheds cover two-fifths of the earth's surface, several hundred of them crossing international boundaries. The binge is slowing now, mostly because there are few rivers left worth damming, and because the opposition, while beaten back by the World Bank, is still vigorous.

The Campaign against Dams

The campaign for, and then against, the World Commission on Dams is the best way of understanding the complicated politics of dams, their funding, and the even more complicated issue of corporate involvement. Major dams have almost always been the creatures of national or state governments, but they have almost always been built by consortia of private water or construction companies. Their financing is often opaque, their real cost often obscured, their benefits often exaggerated, and their ecological damage too often left unexamined.

The activist campaigns against dams that arose through the 1980s were essentially a glomming together of two hitherto unconnected ideas. The first, a product of the environmental movement in its more purist incarnation, was the notion that *any* interference with natural water systems, even flood-control dikes, was damaging and should be contested. A raft of criticisms of dams accumulated, ranging from interference with fish habitats, to salinity caused by the consequent irrigation, to the risk of dam breaches. Most of these criticisms were valid. The second idea was developed by a different set of NGOs, those who called themselves social justice activists, and who were, without exception, anti-globalization and often anti-market. These concentrated on the displacement of people, usually the poor, caused by the construction of large dams. They were able to point to multiple examples, mostly from the early decades of dam building, where local populations had been uprooted, their livelihoods lost, with little or no compensation contemplated or given. This has now been widely acknowledged, and even most pro-dam interests agree that major "interventions" like dams create winners and losers and have a big impact on the economy, on the environment, and on specific groups of people who deserve to be, but often aren't, consulted. It is also generally agreed that the losers, referred to in the careful prose of international financial documents as people in "unmanaged resettlements," have been generally given short shrift.

The catalyst for larger action was the campaign against World Bank funding for the Sardar Sarovar dam on India's Narmada River. This dam, an integral part of India's determination to build what Nehru once called "the cathedrals of modern India," was to be huge, at the time the second largest in the world. It would provide irrigation water for nearly half a million people and enough power for a dozen major cities. It would also submerge a number of fishing villages whose residents were not consulted — they were just told to move: move or drown. One of the villages was a hamlet called Manni Belli.

In June 1994, a concerted NGO effort was launched to bring opprobrium to the World Bank on the occasion of its fiftieth anniversary. The International Rivers network, already a force in the anti-dam movement, teamed up with local activists to issue the Manni Belli Declaration that, among other things, demanded a moratorium on World Bank funding for dams, anywhere. Tactically, the focus on a few hapless villagers, instead of the millions who were clamouring for more reliable electricity, was a success. The villagers, or more accurately, the NGOs operating in their name, called in the prominent Indian social activist Medha Patkar to help. By then the dam was half built, and there wasn't much left of Manni Belli: only the tip of its temple roof was showing above the rising water. That didn't matter. Medha Patkar and her organizers pushed for an immediate stop-work order to the whole thing. In March 1998, a court order halted construction of the dam for a "review period" of three years.

The builders were furious: the millions that had been spent would be wasted if the dam was never completed — there are few things more useless, and unusable, than an unfinished billion-dollar dam. The protesting villagers of Manni Belli who had previously been ignored were now roughed up by hired security thugs, and several of the village women were raped. Patkar was herself beaten and jailed. That was a big tactical mistake.

Even before this, the World Bank, which had recently lost the throw-a-billion-here, throw-a-billion-there muscularity it had expressed under the technocratic Robert McNamara, had been cautiously shifting its focus from the financing of infrastructure megaprojects to something approaching the Millennium Development Goals — financing projects for education, women's empowerment, health, water and sanitation services, and small-is-beautiful projects on scales of small to tiny. The new World Bank president, James Wolfensohn, was trying to navigate between his technocratic traditionalists and the heat he was feeling in the United States from the anti-globalists and was, so far, having only moderate luck. The bank was "owned" by the 180-plus countries that comprised it, most of them intent on development and still impervious to the environmental and social justice activism that was becoming so effective in rich countries. The United States, though, was not impervious. With American politicians beginning to feel the heat — this was before the Tea Party and the rise of Republican anti-government activism — it was becoming harder to justify what seemed to be rough treatment of indigenous peoples, even if the governments of those people, in most cases democratically elected, were the ones doing the roughing. The change was swift enough. The World Bank started as an infrastructure bank, but by the late 1990s, only 5 percent of the bank's lending was still for infrastructure. The bank was essentially out of the dam business.

The Sardar Sarovar fracas greatly spooked the bank's leadership. It did, in fact, yank its financing for the project, and the bank's Operations Evaluations Department was ordered to do a review of the fifty most recent dams built with bank financing. In another tactical mistake, it informed the leading NGO in the Sardar Sarovar fight, the International Union for Conservation of Nature, that the study was underway. The union's leadership immediately interpreted this as a capitulation to their earlier demand for the cessation of dam building, and as an offer to provide the draft text to them for evaluation and criticism.

In September 1996, the completed review was circulated among senior bank staff, but as the International Rivers campaigns director Patrick McCully put it in a press release, "The OED failed to meet its commitment to circulate drafts to NGOs for comment, and the full review was never published." Never mind that no such commitment had ever been made. Only a redacted report was publicly released, which concluded that "the benefits of large dams far outweighed their costs," and that because "thirty-seven of the large dams in this review (74 percent) are acceptable or potentially acceptable, suggests that, overall, most large dams were justified." To the anti-dam people, this clearly wouldn't do.

Still, a meeting was arranged among World Bank officials, dam builders and dam haters. The symposium met in in Gland, Switzerland, in 1997, and was the creature of a senior World Bank executive, John Briscoe, as was the creation of the World Commission on Dams that followed.

At first, the meeting seemed to go nowhere. After its first sessions, International Rivers' McCully issued a press release saying that "participants at the workshop largely agreed that [the bank's study] was based on inadequate data and flawed methodology." This wasn't actually true. Only a few participants — the anti-dam ones — had agreed to any such thing. Others, like the Egyptian scientist Aly Shady, then president of the International Water Resources Association and president of the International Commission on Irrigation and Drainage, asserted that "existing and new dams will continue to play a major role in the [water] management system." Despite this inauspicious start, everyone attending agreed upon Briscoe's prodding to set up a World Commission on Dams, the WCD.

Briscoe's, and the World Bank's, position on the new creation was a calculated mix of co-option (bring the opponents to the table), of genuine recognition that dams did have problems, and of the slightly innocent assumption that a fact-based and objective report would conclude that dams should continue to be built. As Briscoe put it, "The substantive reason [for creating the WCD] is obvious — major interventions like dams create winners and losers, and have major impacts on the economy, on specific groups of people and on the environment. There are many cases where large dams have been the foundation for economic development and well-being of whole countries or regions. No arid country has become rich without extensive investment in water-retaining dams. And no mountainous country has become rich without tapping most of its hydroelectric potential." Then he repeated what many critics had already said, that "these benefits were so obvious that for decades too little attention was given to the negative impacts of dams and, particularly, to the devastating effects which unmanaged resettlement had on millions of poor people. While there are many cases [China, Brazil, Colombia, Senegal, Japan, India, Pakistan, and Canada] in which resettled people have become, as they should, the first beneficiaries of dams, there are too many other cases (more frequent in the early decades of major dam building in the developing world, but persisting in some countries still) where the plight of affected people has been given insufficient attention."[2]

Briscoe therefore hoped a WCD report would recognize the problems and provide ways for dealing with them, clearing the air so that matters could move forward productively. A commission that included among its members representatives of financing agencies and the dam industry would be unlikely to produce a one-note report, he felt. And if the anti-dam people signed off on it, so much the better. The commission was "heavily stacked with anti-dam activists," as he put it later, but "it was my judgment that they would be counterbalanced by some excellent professionals and politicians who had a broader view of the issue. I also believed — incorrectly, it turned out — that the secretary general, a person of great capability, would play a moderating, balancing role."

The position of the anti-dam people was considerably more uncompromising and, in the aftermath, activists like Patrick McCully cheerfully admitted it. I was led to McCully's 2001 summation of the process that appeared in the *American University International Law Review* by Briscoe, who called it the only account he had read that accurately reflected how the process actually worked.[3]

Whereas Briscoe believed that a proper review of dams would be cautiously pro-dam, McCully believed that a "thorough and unbiased review of the actual impacts of dams" would show them to be overwhelmingly negative and portray them as the horrors they were. As to why the dam activists agreed to join a commission that included dam promoters and builders, it is hard not to agree with Briscoe that McCully's analysis was "breathtakingly audacious."

First, McCully described how to use a process apparently set up to find common ground to further the interests of only one party; in his formulation, "to adopt Clausewitz's famous phrase, the WCD was a mere [continuation] of the anti-dam struggle by other means." He recognized that a "multi-stakeholder commission" would be unlikely to take a firm no-dams stance, but he and his colleagues would see to it that "such a commission could set strict criteria for future dams, that, if followed, would prevent most destructive dams from going forward, promote better alternatives, and help promote recognition of the need for reparations for past damage due to dam construction."

His second tactic was how to sideline possible critics:

The largely successful efforts of Commissioners and staff to keep the World Bank's influence at a minimum were also critical to the WCD's success. Several pro-dam governments attacked the process at one time or another, but their efforts were never coordinated or effective. The Chinese and Indian governments both withdrew from the process after initially welcoming it, and, in doing so, only ensured that they could do little to influence its outcome.... The exclusion of governments from substantive power in the process was also vital. Had the governments of leading dam-building nations like Brazil, China, India, Japan or Turkey formed an organized bloc within the Reference Group, it is almost certain that their coalition would have destroyed the Commission's potential to issue a progressive report.

Thus: work within the process to sideline governments and bodies like the World Bank so that "organized civil society" could assure that its own recommendations were successful.

What *were* these recommendations? What did the WCD report actually say? Much of it was anodyne, as McCully had predicted, including bland generalities such as "shared values." In fact, at first it seemed something of a bust because no one quite understood what it was recommending. It even came under assault from some of the anti-dam activists, who denounced it as resolutely refusing to take sides. But buried deep in the report were a series of guidelines for the construction and financing of future dams. Were these guidelines ever to become mandatory — if governments and financing agencies actually adopted them — it is likely that no more large dams would be built, anywhere. And that the word "compliance" kept popping up throughout the report (Briscoe counted sixty instances of the word in just one chapter) suggested that everyone (lenders and borrowers) would have to comply — the "guidelines" would be mandatory. The activists knew very well that "comply with guidelines" would be, to the World Bank staff, an absolute. Comply or don't comply: it was a yes or no idea, a zero-tolerance concept.

Most of the guidelines were unexceptional. But a couple of them spelled trouble for the construction of future dams. For example, a government's right to exercise eminent domain (even with compensation) would be replaced by a guideline that would require "free, prior, and informed consent" for all indigenous peoples, thus giving them a de facto veto over construction. Another guideline said that international financial agencies (aka the World Bank) would be *required* to withhold all future funding until "legacy issues" on *all* prior dams had been resolved (my emphasis). Third, all countries that were not in "good faith dialogue on shared rivers" with all their neighbours would be denied funding from development banks for any purpose in any way related to water.

You can see what McCully means: if the bank adopted those guidelines as prescriptive — no more dams.

In the end, though, it was a pyrrhic victory, and short-lived. When the WCD report came out (strategically leaked to anti-dam activists), even a largely sympathetic World Bank president Jim Wolfensohn blinked. "The critical test," he said at the report's launch, "will be whether our borrowing countries and project financiers accept the recommendations of the commission and want to build on them."[4]

That "whether" was trouble for the McCully crowd. Mark Halle, a veteran Canadian environmental campaigner, suggested that "the Bank had said it would follow WCD's recommendations whatever they were" and that "NGOs are convinced that the bank has come under strong pressure from commercial concerns not to abide by the WCD report."[5] After China, Brazil, Nepal, Turkey, and India all declared that they refused to accept the guidelines, Briscoe and his colleagues were accused of strong-arming these governments, a fairly risible idea. ("Have you ever tried strong-arming the government of China?" Briscoe asked afterward.)

Essentially what happened was that the so-called MICs, or middle income countries, balked at being told by citizens of countries that had completed their infrastructure programs decades earlier that they were not to follow suit. In the aftermath, Briscoe became the lead author in the World Bank's Water Sector Strategy, which kindly acknowledged the work of the WCD but sidelined the guidelines for good. By 2014, the World Bank itself was in turmoil. Its new president, Jim Yong Kim, was radically restructuring the bank, diminishing its country and regional executives, and centralizing power at headquarters. Under this restructuring, fourteen "thematic global practices" were listed as the bank's priorities, and water was one. Some five hundred employees were laid-off, leading to the bizarre sight of a mass demonstration of disgruntled besuited bankers inside the headquarters atrium. The president was derided as an outsider who didn't understand the institution he was running (he was a former health care expert and president of Dartmouth College in New Hampshire). By this time, the bank was financing only a few dams worldwide. Its water priorities were settling elsewhere. In this same year, the Chinese were financing more than two hundred dams outside their own country. World Bank backing represented less than 2 percent of projected investment.

But the Sardar Sarovar dam story has a coda. Early in 2014, just a few weeks after Narendra Modi was elected India's prime minister, the Narmada Control Authority was given permission to raise the dam's level by seventeen metres (from 121.9 to 138.7), making it the second highest dam in the world after Grand Coulee in the United States. The new height would increase hydro generation by 40 percent, and allow irrigation of a further "6.8 lakh hectares" (one lakh is one hundred thousand). It would also affect another two hundred thousand people, drowning the densely populated villages and farmland in the Nimad area of Madhya Pradesh.

None of this was surprising. The increase would benefit, above all, Gujarat State, as well as neighbouring Rajasthan. Before becoming prime minister, Modi was chief minister of Gujarat and had regularly demanded the increase from the central government.

Also unsurprising was the protest by Medha Patkar, still the head of the Narmada Bachao Andolan, the Indian anti-dam group. "The government, she declared, "has neither given us any hearing nor has it taken any time nor made any attempt to know the ground reality before deciding on the dam's height." Greenpeace chimed in, calling the decision "antithetical to the government's promise of inclusive development. With this, the BJP government in Gujarat has served a blow to the Narmada Bachao Andolan and all of us who stand in solidarity with the grave impact of the impacted communities and the environment."[6]

A little more surprising was the political reaction, much savvier than it had been in the earlier go-rounds — no beatings or jailings, just a carefully worded statement from India's water minister, Uma Bharti: "The decision was taken [after consultation with] the ministry of social justice [on the] empowerment of the rehabilitation of displaced people. Social justice... officials are 100 percent satisfied [of the measures being taken by the affected states]."[7]

One hundred percent is no doubt overstating the case, by a considerable margin. But there it rests.

Do Dams Still Make Sense?

Setting the politics of anti-privatization and anti-globalization aside, do massive dams still make sense? This is difficult to assess, given the extravagant claims being made on both sides. The anti-dam people have made their objections admirably clear: dams are really just "brute-force, Industrial Age artifacts that rarely deliver what they promise," as the writer Jacques Leslie put it in 2014.[8] And dam proponents are just as likely to oversell on the other side, with dams being trumpeted as a ticket to national independence and prosperity.

Early in 2014, a team of Oxford researchers, most of them dam skeptics, weighed in with some numbers. They reviewed 245 dams over fifteen metres high built between 1934 and 2007 and found that virtually all of them ran over their approved budget, by almost 100 percent on average, and they took an average of eight years to build, far longer than their proponents had claimed.[9] Dam projects are so huge, they concluded, that cost overruns became contributors to the debt crises in Turkey, Brazil, Mexico, and the former Yugoslavia. To some degree, these cost overruns are simply a fact of modern life.

Megaprojects cost so much and their benefits are often so elusive that proponents almost always lowball the numbers just to get them off the ground, the modern Olympics being a case in point. Further, while the completed hydroelectric dams did provide power, in almost every case it was not nearly as much as their proponents had promised.

But do the Oxford numbers mean that dams are not economically viable? Anti-dam people ask whether fragile economies can afford the tremendous outlays dams require. Besides, foreign companies from richer economies are almost always required to build the things and, they argue, that much of the investment flows back out of the country as a result.

Pro-dam people suggest that dams give economies stability and resilience, and are one of the underpinnings of robust economic growth. Besides, fragile economies have shallow capital markets, and bringing in foreign investment is the proper way to stimulate growth. That is also the classic argument for allowing the privatization of water supply.

An interesting case study to this point is the recently built, but almost-scuttled, dam at the Bujagali Falls on the Nile River, a little below the Ugandan town of Jinja on the shores of Lake Victoria.

When I first went to Jinja more than a decade ago, I drove down to Cliff Road (there really isn't a cliff) to peer at the small plaque marking the Source of the Nile, as it is called hereabouts (a claim vigorously disputed by, among other claimants, Burundi, where there is a small village defiantly named Le Source de Nil). In fact, there seem to be many claimants for ultimate headwater, and Uganda itself now tactfully claims only to be the source of the "Nile proper." When I was there, the water from Lake Victoria poured prettily into its new channel, but there was otherwise not much to see. A busload of German tourists was snapping away, and three Russians with fishing poles were staring morosely at their empty bait buckets. Just downstream, however, out of sight from Jinja were the Bujagali Falls, now obliterated, to the fury of the International Rivers people, by the Bujagali power station.

The Bujagali project was proposed by Uganda's president, Yoweri Museveni, to help overcome his country's chronic lack of energy. Adding 250 megawatts to the Ugandan grid would not only help extend that grid but also substitute for the ruinous costs of diesel generation, to that point the only alternative. The 250-megawatt capacity sounds trivial by the standards of, say, California (where the International Rivers people are based), but it would nevertheless bring electricity for the first time to towns and villages that needed it. The World Bank was involved from about 2003, but its presence faded after vociferous opposition, later returning in a more limited way. The financing was complicated, but typical of such projects. Besides the World Bank, other investors came in and bailed out, sometimes more than once. The consortium that finally brought the project to fruition was under the umbrella of Bujagali Energy, which selected an Italian dam-building company, Salini, for construction. Bujagali Energy was partly funded by Industrial Promotion Services (Kenya) and SG Bujagali Holdings, an "affiliate" of Sithe Global Power of the United States, itself a subsidiary of the Blackstone Group, a private equity and investment fund based in New York. A senior partner in Bujagali Energy was the Aga Khan Fund for Economic Development.

Sebastian Mallaby, in his wildly entertaining book *The World's Banker*, takes up the story: "Western NGOs were in revolt: the International Rivers Network...proclaimed that the Ugandan environmental movement was outraged at the likely damage to the Bujagali waterfalls, and that the poor people near the site would be uprooted from their land and livelihood." The Swedish Society for Nature Conservation was similarly miffed and offered its support. Then, his own bias peeping out, Mallaby suggested that "the activists' resistance had tied up the Bank for several years, delaying a project that would get electricity to clinics and schoolrooms that lacked lights and to industries whose productivity was wrecked by a lack of reliable energy."[10] As it turned out, the outraged Ugandan environmentalists numbered fewer than two dozen, and the villagers, when Mallaby consulted them, were delighted, not angry, to be included — the only angry ones were those who lived just beyond the periphery of the new dam and who would have loved to have been expropriated for compensation.

This was a pristine example of the moral hazard dogging development, in which unelected people who have steady jobs and reliable energy and plenty of money and as much food as they need take on the burden of deciding for those who have none of those things what is good for them and their society. In the end, the Bujagali power system, the third largest in the country, was inaugurated in October 2012 by Museveni and the Aga Khan. World Bank officials were there, but not in the front row.

The Benefits of Dams

The first major benefit of dams is the generation of hydroelectric power. It remains true today that almost 80 percent of the world's renewable energy comes from hydro, with the others (photovoltaics, wind, and the rest) at just a few percentage points. Growing fast, yes, but from a tiny base. For the foreseeable future, only hydro will be able to produce power on the scale of fossil fuels, more than a dozen times the capacity of all other renewables, including solar, wind, geothermal, tidal power, and waste incineration. In fact, hydro is the second biggest producer of the world's electricity, 16.2 percent of the global total in 2012. Rivers are ubiquitous, and some form of hydro power is developed in at least 160 countries. Four countries in the world generate *all* their electricity via hydro, and fifteen others reached at least 90 percent.[11]

The secondary benefit of hydro power is that, once built, it requires no fuel, so a gigawatt produced saves not only a gigatonne of emissions but also all the energy that would have gone into mining the fossil fuel in the first place. And, as Denmark and Norway have amply proved, large reservoirs can act as storage devices for intermittent alternative energies that can then be tapped at will — it is much easier to spin up or down a hydro plant than a coal generator.

As always, Colorado's Hoover dam is everyone's favourite example. Not only was it completed before its deadline and came in $15 million under budget, but the cheap power it produced is widely credited with jump-starting the American economy as it emerged from the Depression. The same is true elsewhere. India's economy has been greatly helped by the cheap (and stable) power that dams produce, which is one reason the Sardar Sarovar dam was pushed to completion. Laos and Vietnam are both dam creators that are now electricity exporters — hydro is a large cash crop for both countries. Little Bhutan is a dam builder too, and Nepal is heading the same way. Brazil has indulged in a binge of new dams, some of them, to the fury of environmentalists, in the Amazon basin — Brazil remains resolutely pro-dam. Canada too. Without Canada's hydro resources, the country's not-very-admirable greenhouse gas emissions would be considerably worse. In the 1970s and 1980s, the government of Quebec, through its state-owned Hydro Québec (the world's largest hydro power, generating more than thirty-six gigawatts for its customers, most of them far to the south), built a veritable cascade of dams in the Canadian northlands, the austere home of the nomadic Cree people. No fewer than eight dams were constructed in the James Bay area, and a dozen others are scattered through the north, some of them immense: the overflow spillway of the Robert-Bourassa dam, until China's Three Gorges the world's largest, can take twice the volume of the St. Lawrence River. The neighbouring province of Newfoundland and Labrador has put up the Churchill Falls barrages, a series of dikes rather than a single dam but having the same effect on the environment, creating a reservoir so large that it alone generates 1 percent of the world's total hydroelectric supply. A new series of reservoirs is being built, and an undersea cable to Nova Scotia will siphon off some of its power and pass the rest through to the insatiable American market.

Some poorer countries with massive hydropower potential, such as Nepal (where anti-dam campaigners are active), currently exploit only about 1 to 10 percent of their hydro potential, whereas on average, rich countries have exploited somewhere around 70 percent. The International Hydropower Association points out that Europe has exploited around three-quarters of its potential sites — the rest are either technically difficult or in scenic areas beloved of national tourist boards. So there is plenty of room for expansion.

Bhutan, for example, famous as the first country to try to replace gross national product with gross national happiness, was finding at the end of the first decade of this century that paying to get there was becoming problematic. Money can't buy happiness — this it knew. But without money, happiness seemed elusive, and this it was still learning. One way was to make money by selling surplus hydroelectric power to India — Bhutan's domestic needs amounted to only 300 megawatts of generating capacity, but the dams it had built yielded 1,480, plenty to spare. In 2013, the little country declared that it would build a new array of dams that would increase its capacity to a substantial 10,000 megawatts. The anti-dam people contested the whole notion, but as of 2014 seemed to be losing.[12]

The James Bay Project in Canada proved that major plants can be sited a very long way from customers, and in some very hostile territory, which implies that many of the Siberian rivers are themselves exploitable. Dams there will almost certainly be built because costs are reasonable, producing power in the same price range as coal. With enough investment, and if local and environmentalist outrage are sidetracked, hydro could be scaled up to produce two and a half terawatts of potential electricity, a respectable proportion of global energy demand — more, in fact, than the total global electricity demand in 2012, and a good piece of the world's total energy use, which is fifteen terawatts. Hydropower will double in China in the next twenty years, and triple in India and Africa, according to International Energy Agency projections.

The second rationale for constructing large dams is for water storage — in wet periods for use in dry ones, for irrigation, and for direct human consumption. I have an inbuilt bias to appreciate this argument, for my grandfather's farm in an arid part of South Africa would not have been viable without the earthen dam he built to contain the sparse rain that did fall, often in torrents. The same thing is true of massive dams. Again, the Hoover dam is an obvious example; the Colorado reservoirs, even as they reached historic low levels in 2015, provide useful water for much of the parched Southwest, including the city of Las Vegas, that could hardly exist without it.

This kind of storage is not necessary where the rains are plentiful, such as Maine, for example, and the American Pacific Northwest, where they are actually decommissioning dams that seem to have outlived their usefulness. Still, most rich countries, as national averages, have around 5,000 cubic metres of stored water for every citizen, poor countries much less: India and Pakistan have less than 150 cubic metres of storage, most of Africa less than that. Pakistan can store only thirty days of the average flow of the Indus River, for example, whereas Australia's Murray-Darling system can store more than a thousand days'. This means that such countries remain much more vulnerable to droughts and to unexpected changes in rainfall patterns than rich countries, exaggerating existing inequalities.

Dams can also mitigate the effects of climate change. To take just one example, and staying with Pakistan, some of the water in the Indus River comes from glacial melt. If these glaciers retreat, as climate forecasts predict, does this mean the water is lost? No, the water will still fall, but more of it will fall as rain and will no longer be released slowly in the warmer months. Instead, it will run through to the sea. More storage in dams would ameliorate the change and act as a buffer in the dry season. The same thing has happened in hundreds of places around the world. The Columbia River that flows from Canada into the United States is a rich-world example: dams provide a net increase in the availability of water in dry periods and flood protection in wet times. The building of a series of large dams in the little landlocked country of Lesotho has provided reliable water to South Africa's Vaal River system on which the metropolis of Johannesburg depends, and provides poor Lesotho with some much-needed foreign currency.

It is precisely these two attributes — that dams generate clean energy and that dams store water for bad times — that make the visceral opposition to dams feel so old-fashioned. I liked the reaction of the reviewers for the website Grist, who in 2014 were commenting on a movie called *Night Moves* (in which two young greenies decide to blow up a dam in Oregon):

> One of the most annoying things about this movie … is that it totally misses the zeitgeist of today's environmental activism. In this century, what drives people out to the streets and even encourages some to risk arrest is climate change and dirty energy. The environmental radicals in this movie don't make a single mention of climate change. They want to blow up a dam because it hurts salmon. What is this, 1975? That's when the idea of blowing up a dam was fresh, when Edward Abbey wrote his classic novel, *The Monkey Wrench Gang*.[13]

The third attribute of dams is flood control, and here the evidence of benefits is ambiguous.

China, as always, provides a good example. One of the reasons for building the Three Gorges dam in China was to ameliorate flooding on the Yangtze, flooding that had killed more than 100,000 in 1911, 140,000 in 1931, 142,000 in 1935, and 30,000 in 1954. In 1998, a major catastrophe was narrowly averted when the riverbanks overflowed and threatened retaining dikes. Had they broken, hundreds of thousands more would have died, and several major industrial cities would have been inundated. As it was, more than 2,000 people drowned, 13.8 million were driven from their homes, 2.9 million houses were destroyed, and 9 million hectares of crops ruined (3 percent of the national total). The floods affected an area inhabited by more than 240 million people, a fifth of China's considerable population.

There are many reasons the Yangtze floods so often. The main natural one is the river's length: it is the longest river in Asia, stretching sixty-three hundred kilometres from its source in China's northwest Qinghai Province to its mouth at Shanghai on the East China Sea. The rainy season is May and June in areas south of the Yangtze, and

July and August to its north, so the wet season is effectively doubled every year. There are human-caused reasons too. Massive deforestation upstream has removed the natural protections that once cushioned the flow. By some estimates, the Yangtze basin, home to four hundred million people, has lost 85 percent of its original forest cover. "The forests that once absorbed and held huge quantities of monsoon rainfall, which could then slowly percolate into the ground, are now largely gone," said then head of the Worldwatch Institute Lester Brown shortly after the most recent disaster.

Three kinds of flooding happen along the river: in the upper reaches, in the lower reaches, and river-long floods. The dam can control only the third type, though it is fair to say that the floods that do still happen upstream of the dam have been diminished in their downstream severity.

This is true of dams elsewhere. They help with flooding. They are not a cure.

The downsides of dams are easy to see. Even their builders acknowledge them.

The careless displacement of populations is the most obvious one, as the World Bank's Briscoe has acknowledged. When Stalin built the series of dams on the Volga River called the Volga cascade, the stepped reservoirs he created submerged hundreds of towns and villages. In typical Soviet fashion, many of the residents were not apprised of what was happening and had to scramble to avoid the rising waters. And when the Kariba dam on the Zambezi River in southern Africa was built in the 1950s, Lake Kariba, the reservoir formed by the dam, became nearly three hundred kilometres long and some thirty wide, one hundred metres at its deepest point, with lush shoreline and dozens of islands. These islands became famous in the 1950s as a rescue operation called Operation Noah issued a worldwide alert for the thousands of stranded animals that would otherwise have drowned (airlifts for stranded elephants and giraffes became a staple of nightly newscasts). Independence movements at the time — for these were still colonies, white-ruled under the short-lived and widely hated Federation of Rhodesia and Nyasaland — pointed out snidely that more concern had been expressed in Europe and the Americas for elephants and leopards than for the fifty thousand Tonga-speaking people who'd been forced to evacuate without an airlift. And in 2014, more than fifty years after the Kariba dam was built, Tonga villages still have no access to electricity.

In another example, Norway is often trotted out as a paragon of green — much of its energy comes from dozens of often small hydro dams — but the country's state-owned energy company, Statkraft, has been financing a series of controversial dams in Laos, India, and Borneo. One of them, a joint-venture dam in Laos, displaced forty-eight hundred people and has been causing flooding, erosion, and loss of fisheries on which thousands depend.[14] Another Statkraft project, in Sarawak, displaced ten thousand people and caused thousands of hectares of tropical forests to be cut down.

The World Commission on Dams report cites several other instances of such displacements. In Mexico, the Papaloapan River Commission set fire to the houses of indigenous Mazatecs who refused to move for the Miguel Alemán dam. In 1978, police killed four people when they fired at an anti-resettlement rally at Chandil dam in the state of Bihar, in India. In Nigeria in April 1980, police fired at people blocking roads in protest against the Bakolori dam. And in 1985, 376 Maya Achi Indians, most of them women and children, were murdered in the course of clearing the area to be submerged by the Chixoy dam in Guatemala.

Of course, all this was not the fault of dams. It was the consequence of bad policy, usually by governments careless of their own citizens' welfare. It could easily have been avoided and the dams still built. One good outcome of the anti-dam crusade has been that their builders and funders, including the World Bank, are now much more scrupulous about displacements than they were. The dams in Lesotho, the Lesotho Highlands Water Project, and the dam at Jinja, on the Nile in Uganda, reveal this new face of dam construction.

Even so, problems persist, and in March 2015, the World Bank issued a sheepish *mea culpa*, acknowledging that it had repeatedly failed to follow its own rules regarding evictions and displacements. "We took a hard look at ourselves on resettlement and what we found cause me deep concern," the bank's president, Jim Yong Kim, put it. He admitted that the bank didn't know how many people its projects had displaced, or what had happened to them. He promised to do better.[15]

And then there is sediment. Dams — all dams — silt up, for dams have a natural lifespan depending on the silt level of the rivers they contain. In certain rocky areas with poor soil (like Lesotho), dams will last for a long time, perhaps for a thousand years, though no one knows for sure. In other cases, like the Nile at Aswan, rivers that once ran brown with silt are now running clear downstream from their dams. By some calculations, somewhere between 0.5 and 1 percent of dam capacity worldwide is lost through sedimentation every couple of years. In some cases, heavy sediment can erode turbines and block intakes.

Nor are dams immune to collapse. By one measure, dam collapses have killed more people than earthquakes or volcanoes. Dams are better built now, but they will never be completely safe.

China's Banqiao dam, on the Ru River, is a tragic demonstration of what I mean. Built in the 1950s to generate electricity but with flood control as its main objective — severe floods had occurred in the basin several times in the 1940s and 1950s — the dam was supposed to be "over-engineered" for safety: it would withstand one-in-a-thousand-year rainfalls, over three hundred millimetres a day. Alas, in 1975, Typhoon Nina collided with a cold front and dumped more than a year's rainfall in less than a day. At one point, the rain was coming down in solid sheets, 189 millimetres an hour, over 1,000 in a single day and more than three times the extraordinary event the dam was built to withstand. When the wall finally burst, it sent 500 million cubic metres of water plunging down the river. All in all, sixty-two dams burst in sequence, and a massive wave ten kilometres wide and more than six metres high with fifteen billion tonnes of water behind it roared down the valley at sixty kilometres an hour. An area fifty-five kilometres long and fifteen wide was completely wiped clear of living creatures, vegetation, and buildings. A new lake of 212,000 square kilometres appeared overnight. Seven county seats were inundated. Some 26,000 people died immediately, and another 145,000 in the days to follow. Nearly six million buildings collapsed. Eleven million residents were affected. Chinese government reports acknowledge that this wasn't the only dam failure that catastrophic summer. Altogether, dam collapses killed a quarter of a million people and brought famine and disease to more than eleven million others.[16]

Other problems with dams are waterlogging and salinity of surrounding water tables (caused mostly, it is true, by the increased irrigation that dams encourage and not by the dams directly), the destruction of downstream fisheries by the blockage of nutrients (though in some cases, a new and sustainable fishery has been created by dams, such as Lake Kariba), and greenhouse gas emissions, mostly methane, caused by the submerged rotting vegetation. There have been claims, widely disputed, that gross emissions from reservoirs cause "almost a quarter of human-caused methane emissions and up to 4 percent of global warming." This sounds alarming, except that human-caused methane emissions are dwarfed by natural ones. In any case, the "peer-reviewed study" from which these alarming estimates are drawn was actually making a different case altogether: that the technologies exist for capturing "dam emissions" and turning them into energy-producing fuel — dams could be an energy source, not a villain.[17]

As the World Commission on Dams report itself said, "The key decisions are not about dams as such, but about options for water and energy development. They relate directly to one of the greatest challenges facing the world in this new century — the need to rethink the management of freshwater resources...large dams [have] emerged as one of the most significant and visible tools for the management of water resources."

Deadbeat Dams

Sometimes, and in some places, it makes sense to decommission dams, to tear them down and let the rivers run free. Perhaps they have outlived their usefulness. Perhaps they have silted up. Perhaps the fisheries they are blocking have more economic value than whatever benefits the dams still' hold. Perhaps they are vulnerable to earthquakes, such as California's thirty-two-metre San Clemente dam on the Carmel River, demolished in 2014. Perhaps they just weren't very well built or would be too expensive to bring up to more rigorous modern safety standards. In the United States, about a thousand dams, most of them lower than five metres but some much larger, have been removed in the last century, including ninety-six in the Pacific Northwest. Two are in Washington State, the sixty-four-metre Glines Canyon dam, on Washington's Elwha River, and the thirty-eight-metre Condit dam on the White Salmon River. All these are what Yvon Chouinard, the founder of the high-end sports clothing empire Patagonia and an anti-dam documentarist (*DamNation*), calls "deadbeat dams."

Large Dams a-Building Still
Despite opposition, mega-dams are still being built. Take Brazil. First under the populist president Luiz Inácio Lula da Silva, and more recently under Dilma Rousseff, Brazil has so far proved impervious to vociferous outsider criticism, especially with two massive new hydroelectric schemes in the Amazon basin, the Jirau and Santo Antonio dams on the Madeira River, erected at a cost of $15 billion. These and other dams in the region (at least two more are planned) are central to the country's Growth Acceleration Program, which aims to stimulate growth through huge investments in infrastructure, especially roads and hydroelectric dams. Brazil is doing much of the financing itself, though the World Bank is participating; construction is being done through a consortium of private and state-owned companies.

The downsides of this binge of dam building are mostly obvious, though some effects, such as flooding in Bolivia due to the Jirau backwater, were less so. Migrating fish remain a concern (the region is heavily dependent on the giant catfish for its food supply, and their migrations would be fatally disturbed). Malaria, already endemic in the region, will likely become worse. Nor were the local people consulted, or at least not very much — many are hard to reach, and some remain "uncontacted." Two national elections have been held since the project was announced, and the dam-building party won both times. The only thing slowing down the building binge is the slowing down of Brazil's economy itself — by 2013, the country had fallen into a difficult recession.

India is another energetic dam builder. In 2003, Prime Minister Atal Bihari Vajpayee launched one of the world's most ambitious energy plans, demanding that the country build no fewer than 162 big hydro-power dams by 2025, later amplified to 292. The new projects would generate better than fifty thousand megawatts, half as much again as the country produced from all sources to that point. Almost all the projects would be built in the five northerly Himalayan states — not so surprising when you consider that steep mountain valleys give a fine head of water for electricity generation, and that dozens of rivers rise there, many of them merging to become the mighty Brahmaputra in eastern India. More than 30 of the dams would be built in one state alone, Uttarakhand, which shares borders with China and Nepal. Some rivers would have half a dozen dams spaced no more than ten kilometres apart.

However, in 2013, in a year of the most intense monsoon season ever recorded, flooding swept down the mountains, killing between six thousand and thirty thousand people (the figures are disputed), destroying tens of villages, carrying away almost a thousand kilometres of roads, destroying twenty or so small hydro projects, and seriously damaging ten large ones. Two months after the flooding ended, India's Supreme Court essentially shut down further dam building, pending further analysis and "scientific study." The International Rivers people were exultant: "It shows that nature will strike back if we disregard the ecological limits of fragile regions like the Himalayas through reckless dam building and other infrastructure development. We can only expect such disasters to happen more frequently under a changing climate."[18] Alas, one consequence is that India will have to rely more on burning coal to generate the power it needs.

Nepal uses less than 1 percent of its hydro potential, but that is about to change in a big way, despite the disastrous earthquake of 2014. A cluster of Indian dam builders, among them the giant GMR, are building (or planning) a series of multi-billion-dollar projects that would, if all were completed as planned, plausibly earn Nepal $17 billion over twenty years — this is a country whose GDP in 2014 was $19 billion. Much of the electricity these dams would generate would be exported to India. Indian engineers have suggested that if all Nepal's rivers were thus tapped, they could produce somewhere around forty gigawatts of power, almost a fifth of India's current capacity.

China is still the behemoth of dam builders. In addition to its inventory of nearly nineteen thousand large dams, China's energy plan in 2013 projected the construction of another forty or so that would have twice the hydropower of the United States.[19] Five of Asia's greatest rivers — the Red, the Yangtze, the Irrawaddy, the Salween (called the Nu in China), and the Mekong — start within a few hundred kilometres of each other on the Tibetan Plateau, and China is planning hydro dams on all of them, even the pristine Nu, which flows through a UNESCO World Heritage Site in Yunnan Province, cited as one of the world's most ecologically diverse and fragile places. More than sixty thousand people are likely to be displaced by the Nu dams. Pleas were ignored from, among others, Thailand and Myanmar, both of which depend on the Salween for basic subsistence. China never did sign a 1997 UN water-sharing treaty that would have mandated consultation with affected neighbours, in this case, India, Myanmar, Thailand, Vietnam, Kazakhstan, and Russia. And it is planning a dozen dams on the Mekong (called Lancang in China) that flows for forty-four hundred kilometres through six countries — China, Myanmar, Laos, Thailand, Cambodia, and Vietnam — to the South China Sea, scooping in a population of more than sixty-five million.

Protests from downriver may just be a case of the kettle calling the pot black. Cambodia, Laos, Vietnam, and Thailand are all building, or have announced plans that they will be building, dams across the lower Mekong. At least nine new dams would be in Laos. There is an uproar in the making between dam builders, mostly Chinese and Malaysian companies, and the lower Mekong fishing industry that generates over $4 billion a year. A review by the multinational Mekong River Commission in 2010 asserted that the new Mekong reservoirs would reduce the value of the fishery by half a billion a year, put a hundred fish species at risk, mean the relocation of more than a hundred thousand people, and seriously damage the productive riverbank farming. Only about a quarter of the river's flow would any longer get to the sea, threatening the Mekong Delta and Vietnam's rice paddies.

It is possible to be pro-dam, even if cautiously. But damming the Mekong to provide less than 10 percent of the region's projected electricity demand, set against the consequences outlined above, seems unnecessarily perverse.

And then there is the Congo River dam, an improbable plan to build in an improbable place at improbable cost the greatest hydroelectric project on the planet — more than double the output of China's Three Gorges system, capable (if ever built) of providing enough electricity to supply half of sub-Saharan Africa with reliable power. And yet it might get done. The World Bank has approved a $73.1 million grant — trivial enough, given the project's estimated $80 billion-plus cost, but, together with the African Development Bank's similar grant, a significant marker of feasibility. Enough to get the legal framework in place, in any event.

In 1999, I had spent some time in Brazzaville and Kinshasa, twin Congo capitals on either other side of the river. My notes, later collected into a book, had this to say: "I sat for a while on a rock, staring at the Congo River rapids southwest of the Zairian capital, Kinshasa, watching islands of hyacinths come down from the deep interior of tropical Africa dashing themselves to pieces on the glistening boulders, a strange and exotic sight.... The water had begun more than four thousand kilometres away, near Lubumbashi in Shaba Province, almost on the continental divide, then drifted placidly through Kisangani in Haute Zaire, where the rebels had started their 1997 drive to wrest Zaire [as the Democratic Republic of the Congo was then called] from its aging and ailing dictator. A long and leisurely journey it was; only here, at the rapids, did the river seem in any hurry." The Congo is unusual. Most river rapids are upstream. Congo is unique in that only near the mouth does it drop suddenly, a total of thirty-five metres in a series of small falls and rapids. "This was one of the world's longest rivers, with the third greatest flow, after the Amazon and the Ganges. Every second, between 30,000 cubic metres (July) and 55,000 (December) pass by any one spot. If these Congo rapids were channelled and electrified, someone had calculated, they could supply about a sixth of the world's hydroelectric power. Seso Seko Mobutu, Zaire's unlamented late dictator, had had, in fact, a typically grandiose scheme to provide power for most of Africa by doing precisely that."[20]

In fact, Mobutu, when he wasn't siphoning off money to stash in France, had caused to be built two smaller prototypes, Inga I, in 1972, and Inga II, a decade later. They were built largely to supply energy to nearby aluminum mining and smelting operations. Together, they are supposed to produce around two thousand megawatts, a very long way from the more than forty thousand projected for Inga III, or Grand Inga (some plans say as much as a hundred thousand megawatts). Even so, their actual output is less than a quarter of capacity — graft has "misplaced" much of the money and materials for maintenance, and the people displaced are still waiting for compensation three decades later. This is not atypical. Any construction in the Democratic Republic of the Congo is a locus for corruption, grotesquely low-balled cost estimates, and wildly optimistic completion deadlines. So when you read that construction is likely to start in 2016 and to be completed in 2020, you would be wise to express a little skepticism.

Still, there are serious partners involved. Grand Inga is financed by a public-private partnership and is listed as one of the G20's top ten "exemplary transformational projects." The European Investment Bank has chipped in. Other partners include the New Partnership for Africa's Development, the Southern African Development Community, and Eskom, South Africa's state-owned electrical utility, the largest in Africa, which will provide technical expertise along with France's utility, EDF. In exchange, South Africa, which has a serious supply shortage, has a deal with Congo to get the first twenty-five hundred megawatts of stage one's forty-eight hundred. Three contractors will share the load: Chinese, Spanish, and Canadian-Korean. Hydro Québec was involved in early stages but subsequently dropped out.

The plan is for Grand Inga to be constructed in six phases, with Inga III only the first. Inga III won't close the river but divert part of it to an adjacent valley that runs parallel to the main stream. Later stages will completely flood the Bundi Valley for a twenty-two-thousand-hectare reservoir that will displace large numbers of people and provide a new home for even more billions of noxious dengue fever–carrying mosquitoes and parasites. It is unlikely that the displaced will receive much in the way of benefit or compensation. None of the plans envisage bringing electricity to many of the Congolese people, except those in Kinshasa itself. And ecologists are worried about the loss of the famous Congo plume and its effects on local, regional, and perhaps even the global climate. The feasibility studies that have taken place are all about cost and construction; the environment has received only glancing attention. Environmental opponents have been reduced to hoping that corruption and escalating costs will mean the thing never gets built.

The International Rivers network has pleaded for rethinking. Yet they themselves are beating the same tired old drum: "G-20 leaders should prioritize investments that directly address poor peoples' needs rather than using taxpayer money to pay for huge, high-risk projects whose private sector returns rarely trickle down. In sub-Saharan Africa, where water stress is highest, decentralized infrastructure has a better track record of providing water and electricity than do large storage dams, which can worsen poverty and reduce climate resilience, and are certainly not green."[21] That from Zachary Hurwitz, policy coordinator of International Rivers. There was no sign anyone was listening.

7
What We Don't Know
about Fracking

Somehow, it is reassuring when a scientist quotes a poet to explain his own uncertainty. It is even more reassuring when the scientist occupies a space and professes a profession that would have been unthinkable just a few decades ago, for he is a "contamination hydro-geologist." This means he spends his life examining the causes and effects of large-scale human error. And more reassuring yet when the poet is the ever-gloomy William Blake, and the quote is from his "Proverbs of Hell" (part of *The Marriage of Heaven and Hell*, written somewhere around 1789), for Blake was not just a gloomster but a skeptic to his bones.

The quote is one of Blake's better-known aphorisms: *You never know what is enough until you know what is more than enough.* The scientist John Cherry had just presented a report commissioned by the Canadian government on the merits, hazards, and benefits of hydraulic fracturing, and the Blake quote helpfully summed up his panel's main conclusion: that no matter what point of view you are currently hearing, whether from propagandists of the drilling industry or from environmentalists who oppose fracking in all its forms, the assertions made are almost certainly unjustified. In the year 2014, decades after fracking became widespread, we just don't yet know whether enough is already enough or whether enough will be more than enough pretty soon. All the papers he'd read, Cherry later told a conference audience in Toronto, declared for one side or the other, but none of them ever revealed by what processes their conclusions were drawn.

Cherry's day job is at the University of Guelph, in Ontario, in the Centre for Applied Groundwater Research. He is also chair of the Council of Canadian Academies, self-described as an organization that pulls together scientists and other experts to provide independent policy-related assessments, but which stops short of explicit advice. The particular panel in question, the clumsily but accurately named Expert Panel on Harnessing Science and Technology to Understand the Environmental Impacts of Shale Gas Extraction, had been commissioned by Canada's then environment minister, Peter Kent, whose political bias was such that he no doubt hoped for a resounding pro-fracking conclusion.

Not so. Groundwater contamination? Don't know yet, basic science not yet done. Methane leaks? Probably, but not enough credible science has been done. Contamination of surface or shallow wells? Yes, but from fracking or merely from poorly designed wells? Don't know. Science has not been done, not yet.

In a conversation a few months after the report was released, Cherry told a small audience that if pushed, he would probably opt in favour of fracking, just as he would for nuclear power, "if properly executed." And that was the problem, he said. Fracking is being done without proper monitoring and without any real idea what proper execution would actually be. And, "unlike big hydro plants or nuclear plants, fracking takes place near, and sometimes underneath, ordinary people."[1]

For example, he said, one of the problems involves leaking wells, not just leaking noxious chemicals into the groundwater, as alleged, but allowing methane to seep into well water and into the atmosphere. The oil and gas industry recognizes that leaks are an issue, he said, though their propaganda machine will not acknowledge it. "How to prevent leaks, then? Problem is, no one knows how the gas actually leaks, and how the leaks affect groundwater, if they do. Not one single paper examines this issue," he noted. Russell Gold, the *Wall Street Journal* reporter and author of an exuberant book on fracking called *The Boom*, told the same audience of seeing rainwater puddling around a wellhead in Texas: "You could see bubbles coming up, obviously gas leaking.... But the well owner's solution was to remove the water so there were no longer bubbles to be seen.... Dozens of tests were made around that well, but the source of the leak was never found."

In September 2014, a study of gas leaking from fracked wells was published in the US *Proceedings of the National Academy of Sciences* (*PNAS*), and concluded that yes, drinking water was being contaminated, but no, it was almost certainly not from the fracked deep-level shale but from improperly sealed wells at shallower levels. Cement used to seal the outside of the wells or the steel tubing used to line them was at fault, resulting in gas leaking up the wells, into aquifers, and into the atmosphere.

The report was seized on by the fracking industry, which said it "proved" that fracking was safe. In fact, they went further and asserted that there has never been a "proven instance" of fracking contaminating groundwater. Here's a quote from a company called Cuadrilla, which is trying to get fracking going in England, among other places: "There have been over two million hydraulic fracture treatments carried out globally, the majority in the United States, and from that activity we are not aware of one single verified case of fracturing fluid contaminating aquifers." In this Cuadrilla was just echoing what ExxonMobil's chair, Rex Tillerson, told a congressional panel in 2013: "There have been over a million wells hydraulically fractured in the history of the industry, and there is not one, not one, reported case of a freshwater aquifer having ever been contaminated from hydraulic fracturing. Not one."[2] This careful, misleading phrasing really annoys fracking opponents, who point out, quite correctly, that it is splitting hairs — that if wells on the way down to the fracked area leaked, fracking wasn't off the hook. Drilling was, in their definition, a necessary part of fracking, and they can point to countless fines and penalties assessed on drilling companies for contaminating groundwater with methane and for dumping toxic chemicals into streams.

Cherry, for his part, referred to a paper by the University of Waterloo's Maurice Dusseault, called "Why Oil Wells Leak," which pointed out that cement inevitably deteriorates, even when done properly. "The Germans too," said Cherry, "have found that gas-well cementing does not remain leak proof."

So the assertion by Peter Kent's successor, Leona Aglukkaq, that "shale gas deposits can [therefore] be developed safely, responsibly, and in compliance with the strict environmental policies and regulations in place,"[3] is hokum. She can know nothing of the kind. She may be right, of course, but her assertion is fact-free.

It is useful to recap what fracking really is, to get round this notion that it can be neatly compartmented into drilling and everything else.

In one way, of course, it can. You have to drill to get down to the fracking zone, but you don't have to frack when you get there. This kind of drilling, directly into coal seams or shale beds containing methane gas, has been done for decades. Older wells of this kind used some of the same techniques to shake the gas loose, injecting water under high pressure to crack the shale, allowing gas to seep out. Water wells have done the same thing — the drilled well at my home in Nova Scotia fractured the shale to allow water to pool out, and we still after ten years have to filter the water to get rid of tiny shale fragments.

But fracking as it is currently defined is rather different. The main difference is that it employs horizontal drilling from the original well base, with channels punched up to a kilometre in spoke-like patterns, giving each drilled well a much bigger footprint deep down. Each fracking pad is a major industrial undertaking, with ten to twenty holes on each pad, leaving behind a pocked landscape. Even more controversially, the water that it blasts down the wells to fracture the rock is laced with a witch's brew of abrasive sand and chemicals, many of them proprietary and therefore unknown to regulators, and all of them persistent and liable to leach into water tables. "Flowback water," the chemical-laden stuff that comes to the surface with the gas, has to be disposed of. In some areas it can be pumped back deep underground, but only where the geology permits. In other areas, it must remain on the surface, in massive leaching ponds the size of lakes.

As a business, fracking is transforming the energy landscape, though to what extent remains disputed. By the end of 2014, the United States was drilling about a hundred wells a day and the tempo was still accelerating. The Marcellus shale area of Pennsylvania and New York is claimed to be the number two oil-producing region in the world, after Russia. In British Columbia, somewhere around eleven thousand wells were drilled in 2014 alone. The fracking industry currently uses as much energy to extract the gas as eleven multi-megawatt nuclear power plants.

The boast is that fracked gas will be a bridge technology, weaning electricity generation off coal, and transportation off diesel and gasoline. This too is disputed.

So here are the issues:

- Is the promised size of the transformation real, or is it hype, a short-lived boom before the coming bust?
- Are the vast amounts of water needed for fracking justified? Is the water available? Is there enough, or will fracking divert essential water from other applications?
- Does fracking contaminate groundwater and poison wells? What are the contaminants, and what do we know about them?
- Can wells be made leak proof?
- Does fracking allow the escape of more methane, a much more potent greenhouse gas than carbon dioxide, than acknowledged?
- Does fracking cause earthquakes?

How Big an Industry Is Fracking, Really?

Fracking, both for oil and gas, was supposed to be the game changer. It was going to turn the United States into an energy superpower, a net exporter instead of an import hog, with oil and liquefied natural gas as a global power chip. As of 2012, according to industry sources, slightly more than two and a half million wells had been fracked worldwide, more than a million of those in the United States. The typical well produces a gusher of gas or oil that quickly settles down into a stream. The big question remains: Will that stream produce gas for decades, or quickly run out? Skeptics of Russell Gold's *The Boom* suggest the supply will soon peter out and the boom will be over, unless the industry drills an ever-increasing number of wells in ever-increasing numbers of places.

So far, the shrinkage has been minimal. In fact, although the number of drilling rigs has remained steady for a couple of years, production is still increasing, as is productivity, mostly through more sophisticated extraction techniques. No one really knows whether this will last, or for how long, or how much further each well can be pushed.

In one way, of course, this became moot late in 2014, as gas and oil prices fell through the floor and the drilling rigs fell silent ("Up to their derricks in debt," as the *Economist* put it). But no one expects those rock-bottom prices to persist forever.

What is clear is that early estimates of the potential reserves were vastly overstated. Nothing illustrates this more than the potentials that were publicized for the Monterey shale drilling area south of San Francisco, in California — right on the San Andreas fault, as it happens. In 2011, the Energy Information Administration (EIA) contracted out a study of the Monterey "play," as it is called, to a company called Intek, which seconded research to an analyst named Hitesh Mohan. Recoverable oil and gas, Mohan wrote, amounted to a very substantial 15.4 billion barrels (almost three and a half trillion litres), almost double the productive Bakken shale in Montana and five times larger than Eagle Ford in Texas. Developing the field would add 2.8 million jobs by 2020 and produce a gusher of taxes into state and federal coffers, $24.6 billion worth. As it turns out, however, the study's numbers were based almost entirely on optimistic technical reports from petroleum companies, including Occidental Petroleum, which owned millions of acres of leases in the Monterey area.

Two years later, the EIA, attempting to wipe the egg from its face, published revised estimates, this time based on known geological factors. The new estimates were about 90 percent lower than the original ones. There would not be 15.4 billion barrels of recoverable oil and gas, but 600 million. It wasn't only that the original estimates of the amount of gas were inflated but also that the jumbled geology of the area, sitting as it does at the junction of active tectonic plates, made recovery incrementally more difficult.

The write-down vindicated opponents of Californian fracking, such as the Post Carbon Institute, whose spokesman, the geoscientist David Hughes, wrote that the Monterey formation "was always [a] mythical mother lode puffed up by industry — it never existed."[4]

So what, then, to make of other industry analyses elsewhere in the country?

The Marcellus shale, which underlies parts of New York, Pennsylvania, West Virginia, Ohio, and Maryland, is described on the website energyfromshale.org as "one of the largest shale regions in the United States[,]...estimated to be the second largest natural gas find in the world....The 400-year-old [sic] Marcellus shale region is estimated to contain more than 410 trillion cubic feet of natural gas, and could supply US consumers' energy needs for hundreds of years." Penn State University's Terry Engelder, described in the *Philadelphia Inquirer* as "a leading advocate of shale gas development," was quoted there as putting the reserves at 4,359 trillion cubic feet, a nicely precise number, of which "perhaps 30 percent may be recoverable." The region could produce for perhaps twenty years or more.[5] However, on other petroleum industry websites, Engelder's estimates were 500 trillion cubic feet, which more closely match those of other geologists.

Other shale areas include Eagle Ford, near Dallas, Texas, which best guesses say holds somewhere between 150 and 180 trillion cubic feet, plus twenty-five or so billion barrels of oil. The Barnett shale field, which underlies the Dallas–Fort Worth metro area, has already produced five trillion cubic feet of gas and is expected to yield up forty trillion more — this estimate from the fracking giant Chesapeake Energy, which in 2013 set a record of sorts by becoming the first company to drill a well in the middle of an international airport. The Bakken shale area, which mostly produces oil and not gas, is now pumping a million barrels a day — but it will likely take twenty-five hundred new wells a year to sustain that rate, and it is far from clear how long the reserves will last.

David Hughes, the geoscientist who pooh-poohed Monterey, has produced a substantial study of the fracking industry, sardonically titled *Drill Baby Drill*, after the mantra of the industry's congressional shills. His conclusion, not at all sardonic, was measured: the United States was not going to become the new energy superpower to rival Saudi Arabia. Or Russia, for that matter.

Nor would it become a major exporter. "New technologies of large-scale, multistage, hydraulic fracturing of horizontal wells have allowed previously inaccessible shale gas and tight oil to reverse the long-standing decline of US oil and gas production," he wrote. "This production growth is important and has provided some breathing room. Nevertheless, the projections by pundits and some government agencies that these technologies can provide endless growth heralding a new era of 'energy independence,' in which the United States will become a substantial net exporter of energy, are entirely unwarranted based on the fundamentals. At the end of the day, fossil fuels are finite and these exuberant forecasts will prove to be extremely difficult or impossible to achieve."

The much touted reduction of US oil imports, he pointed out, has been mostly a story of reduced consumption — the recession caused that, not conservation or extra domestic production. As he told the press at the time, "The good news is that supply grows short term, but the bad news is that we may have a very serious supply issue 10-15 years out." Then he came back to Monterey: "[This] was a huge field wiped out with a stroke of a pen. That's like two Bakkens off the table in one fell swoop."

All this hasn't been helped by a Bloomberg News investigative report in October 2014 that found that virtually every fracking company was inflating its shale reserves.[6]

Water Use in Fracking and Its Consequences

Here are some facts:

Fracking a single horizontal well for gas uses an average of 18 million litres of water. These numbers reduce somewhat for oil: 121 million litres in a fracked well for oil. Simple vertical wells use 2.6 million litres for gas, close to 2 million litres for oil.

As techniques get more sophisticated, and more gas is generated from each well, water use goes up. As Russell Gold pointed out to the same small audience addressed by John Cherry, in 2003, a really productive well used 10.6 million litres of water (and 100,250 kilograms of sand); in 2013, the Four Sevens Oil Company drilled a well in Susquehanna County, Pennsylvania, that required some 47 million litres.

In 2013, 70 percent of fracking was for oil, not gas.

In the United States, 56 percent of wells are being drilled in water-stressed areas, 36 percent in areas of severe groundwater depletion. According to the Water Resources Institute definition, "high-stress" means that over 80 percent of the water in those areas has already been allocated, and there is competition for the rest. In Colorado and California, 97 and 96 percent of the wells respectively were in high-stress areas. In Texas, 52 percent.

Over thirty counties have each used four billion litres of water for fracking, roughly equivalent to the daily use of eight million people. Dimmit County, Texas, in the Eagle Ford play, used the largest volume of water for fracking in the United States — about fifteen billion litres. Wells in the Permian basin of Texas use water from aquifers that overlap the Ogallala aquifer itself, and drilling there is slated to double by 2020, from around three thousand new wells a year to six thousand. In Colorado, fracking in the small Denver-Julesburg basin already uses more than the city of Boulder uses in a year.[7]

Three companies — Halliburton, Schlumberger, and Baker Hughes — account for about half the water used nationally. The biggest single water user is Chesapeake Energy, though most of its operations are not in high-water-stress areas.

Russell Gold again: "The two areas of Texas where there is most fracking are also the areas of worst drought. In the Eagle Ford region, there was not enough water even before fracking started." In the freewheeling Texas water market, the Rio Grande has been fully allocated by companies buying up existing water rights. "And since companies have much more money than farmers, they are acquiring all they need. Some cities don't even know this is going on. It is not a fair fight. Water is a local market but the companies are global."

All this just to get oil and gas from the ground. Existing conventional oil drilling, even before fracking started, was using somewhere around two billion gallons a day from lakes, streams, and groundwater, according to US Geological Survey figures. The refining process — to turn the stuff mined into usable fuel — used another two billion or so, mostly for cooling. It is all further complicated by the otherwise sensible notion of building new refineries closer to the source, further stressing the local supply. A new refinery called Hyperion in southeast South Dakota will use 45 million litres a day for processing and cooling, if it gets built — this from an aquifer already stressed by agricultural over-pumping. The plant's wastewater would afterward be dumped into the Missouri River. Unsurprisingly, there is considerable local resistance to the whole thing.

One consequence of the drought is that drilling companies are beginning to look at wastewater (or "produced water," in the jargon of the trade) as a potential asset rather than as a cost. Recycling such water is still only done in trivial amounts, though those amounts are going up. Few companies recycle even half the water they use, and those that do say it costs more than dumping it down "disposal wells," of which there are many in geological areas like Texas, where such wells are plausible. But recycling does cut down on other disposal and trucking costs. And in the entrepreneurial culture of the fracking industry, some are seeing opportunities. Consider, for example, companies with names like Water Rescue Services and Fountain Quail Water Management, that didn't exist half a dozen years ago. Some of them are looking beyond the drilling industry. If enough produced water can be cleaned up sufficiently, it could be used for other purposes, such as municipal water systems and farming, in which case its value would increase substantially. It could even be used to recharge aquifers.[8]

A small curiosity is that an oilfield near Bakersfield, California, the Kern River field, is actually supplying water to the local water district that resells it to farmers — water acquired from the same underground rock formations that contain the oil. The oil and water are separated and the saline water diluted, then pumped ten kilometres to Bakersfield's Cawelo water district.

As the *Economist* reported in 2013, a technique developed by a German-Singaporean company called Memsys for desalting seawater could provide a relatively cheap and effective way of separating fracking water from its contaminants, including salts. The technique is called "vacuum multi-effect membrane distillation," and combines two well-known desalination techniques: distillation and filtration through membranes, called reverse osmosis. The company claims it sharply reduces the energy required for desalination, which if it proves true could make cleaning up fracking fluids much more common.[9]

Other companies are looking to use the brackish water that exists as a by-product of desalination as source water for fracking purposes.

Groundwater Contamination

A "progress report" from the Environmental Protection Agency (EPA) in December 2012 listed more than a thousand chemicals reported to have been used in fracking or detected in fracking wastewater. The EPA used the phrase "reported to have been detected" because it didn't do the analysis itself, instead relying on self-reporting from the nine largest fracking companies, as well as a national database called FracFocus.org, an industry-government fracking chemical disclosure registry. Most of the chemicals were known, and so were their properties. Others, not at all. Nor were their quantities always recorded — FracFocus is far from complete, and its mandate allows companies to avoid disclosing chemicals they consider trade secrets. Indeed, sometimes the companies themselves don't always know what chemicals they are using, having outsourced fracking fluids to third-party contractors. Nor did the EPA report go into the matter of where these chemicals end up, or by what pathways, or how much of the stuff remains underground.[10]

In any case, the EPA, the agency tasked with protecting the environment, is specifically forbidden from regulating the injection of fracking fluids under the Safe Drinking Water Act. It can regulate pretty well everything else to do with water but not this — it's an exception that was passed by Congress at the urging of Dick Cheney, then the vice-president but formerly boss of Halliburton, one of the biggest frackers in the world. This — the hands-off fracking clause — is widely known as the Halliburton Loophole.

Some things are known. One of the drilling companies that does recycle its fracking water, Fasken Oil and Ranch, says the solid-waste residues from cleaning up fracking-produced water includes boron, sulphates, and radioactive metals. All fracking fluids are highly saline, much more so than sea water. Methane is common. Toluene, benzene, and ethylbenzene, all of which affect the human central nervous system, are also used in fracking, though most exposure to these chemicals would come from air pollution, not from water. The toxic solvent 2-butoxyethanol, commonly called 2-BE, is a common ingredient in fracking fluids. This stuff, 2-BE, gave rise to a lawsuit that environmentalists derisively call the "Windex Defense," when a Colorado woman, Laura Amos, sued Encana Corporation blaming a well blowout near her home for the rare tumour she developed. After an "investigation," state regulators suggested to Amos that fracking wasn't to blame, and that if Amos indeed had been exposed to 2-BE, it probably came from Windex, a window cleaner. Encana naturally denied any responsibility but hastily shut the case down by paying Amos a multi-million-dollar settlement and buying her family property. The makers of Windex remained silent throughout the affair.

No one disputes that these chemicals, and many others, are present in fracking fluids. The industry maintains that it is "highly improbable" that they can get into drinking water. Most critics actually agree. But not always: there have been many documented cases of gases bubbling into water wells. Even the industry admits that on occasion fracking fluids do indeed end up in aquifers, but it maintains that it happens where well casings have failed, and that failure rates have dropped significantly as drillers acquire more experience. In an early West Virginia case dating back to 1987, fracking fluids got in a small aquifer whose very existence had not been known, migrating upward through older oil and gas wells that had been abandoned years earlier and plugged with cement, though obviously not well enough, eventually making their way into at least one farmer's drinking water, that of James Parsons, of Jackson County, some 183 metres from the drilling site. As the journalism website Truthout discovered, a US Department of Energy investigation had found about two and a half million abandoned oil and gas wells in the United States at the time. Parsons never made much of a fuss about his contaminated well, and the case was quietly sealed when he reached an agreement with the drillers, Kaiser Exploration and Mining Company, and was paid an unknown sum in compensation.[11]

In July 2014, California state regulators shut down a bunch of wastewater injection wells, fearing they had contaminated nearby groundwater. An EPA report was supposed to be delivered in sixty days, but it wasn't until a year later that the report was finally released, delayed by ferocious industry lobbying. As it turns out, nearly three billion gallons of toxic wastewater had been illegally injected into central California aquifers, and half of the water samples collected showed high levels of dangerous chemicals such as arsenic and thallium, widely used as rat poison. Nine of the eleven sites investigated were illegally dumping chemicals into groundwater. If anyone had thought industry self-policing would work, they were now disabused of that notion.

Best evidence suggests that fracking wells will probably always be susceptible to leakage, but if best practices are used, leaks can be reduced to about half their 2014 values.

Methane Seepage?

Methane, which we weirdly call "natural gas" when we burn it, is the third largest cause of the greenhouse effect, after carbon dioxide and water vapour. It is well known that methane is actually a much more potent greenhouse gas than carbon dioxide, roughly thirty times better at trapping heat in the atmosphere as its bad-boy cousin. It is, on the other hand, short-lived and doesn't stay in the atmosphere for long. But it is a bad thing to let loose unburned.

The fracking process is, after all, an attempt to get at all that methane trapped underground and bring it up for use. It is hardly surprising that some of it leaks. It is difficult enough to prevent liquids escaping from wells and pipes; gas is even harder to corral. As Bob Howarth, a biogeochemist from Cornell University, told Bill McKibben, "It's a hard physical task to keep it from leaking — that was my starting point. Gas is inherently slippery stuff. I've done a lot of gas chromatography over the years, where we compress hydrogen and other gases to run the equipment, and it's just plain impossible to suppress all the leaks. And my wife, who was the supervisor of our little town here, figured out that 20 percent of the town's water was leaking away through various holes. It turns out that's true of most towns. That's because fluids are hard to keep under control, and gases are leakier than water by a large margin."[12]

Howarth published a study in the journal *Climate Change* in May 2011 that concluded that somewhere between 3.6 and 7.9 percent of methane from fracked wells was escaping, a pretty big number. Another study, published in *PNAS* early in 2014, found high levels of methane in the atmosphere over a few wells and nothing very much (or at all) over the majority, which seems to indicate that methane escapes are preventable with proper well casings. The paper also suggested that the higher methane levels might not be due to the fracked wells but to the existence of nearby coal seams, though it was inconclusive on the matter — the fracked wells might have drilled through coal seams on their way down, which would combine the two sources.[13]

Meanwhile, Colorado, alone of the states that allow fracking, has devised a set of rules aimed at reducing methane, written in collaboration with the Environmental Defense Fund. Drillers are now obliged to test for leakages monthly, and if necessary will have to retrofit wells with better valves and casings. The industry has agreed to go along.

Late in 2014, the EPA issued a report called *Waste Not: Common Sense Ways to Reduce Methane Pollution from the Oil and Natural Gas Industry*. The report provocatively suggests that the escaping emissions from existing natural gas wells, compressors, and other equipment could be enough to heat more than six million homes.

Does escaping methane matter from a water perspective? Not really.

Sure, producer Josh Fox's provocative documentary *Gasland* showed the now famous image of a kitchen tap shooting flame, and a newspaper headline, after a recent study (by scientists at Duke University) into fracking, declared that "Scientific Study Links Flammable Drinking Water to Fracking." This brought about the now predictable reaction: the fracking people protesting that it was the well drillers who were responsible, as though it mattered to the hapless homeowner. This happens more often than it should, but that is not very often, and the number of incidents is shrinking as industry starts to do something about it. From a water perspective, fracking fluids are a more serious issue.

Fracking and Earthquakes

Fracking does trigger earthquakes. Cornell University's Katie Keranen took a detailed look at the extravagant number of small quakes near the town of Jones, Oklahoma — 2,547 of them in five years — and found a direct link to fracking. Not so much to the practice itself, but to the vast quantity of wastewater injected far underground into so-called disposal wells. Jones is the home of what Keranen calls "four, modern, high-rate injection wells" that dispose of four million barrels of wastewater a month, and these wells, she found, "impact regional seismicity and increase seismic hazard."[14]

On the plus side, another study by Susan Hough of the US Geological Service found that the quakes were all of low intensity and caused significantly less shaking than naturally caused quakes, perhaps because the very fluid that causes them also lubricates the shifting tectonic plates.[15]

Similar clusters of small quakes have been found at other fracked sites, including in Pennsylvania and Oklahoma. The quakes themselves don't matter from a water perspective, but the poisons pumped into the disposal wells do.

The Egregious Tar Sands

The environmentalist position on the Athabasca tar sands (or oil sands, as its producers prefer) is admirably uncomplicated: the entire project, Alberta's crude-oil engine driver, is wholly evil. In this view, mining these immense reservoirs of bitumen for conversion into usable oil represents one of mankind's largest and ugliest assaults on the natural world. For those who have seen all three, only the West Virginia–Tennessee coal district and possibly Russia's Donets region even begin to compare in their ghastliness. One of the oil sands' largest proprietors, Murray Edwards, chairman of Canadian Natural Resources, told a panel of businessmen and lawyers in Lake Louise in November 2014 (rather glumly, I thought) that the "antis" were not going away anytime soon. The oil industry would just have to get used to it. The environmentalist NGOs, he suggested, really didn't want the oil sands to go away, because they were such a convenient whipping boy.

The producers' cause has not been helped by the obvious ineptitude of their early attempts at favourable publicity. Not just their insistence on squishing the notion that "tar" had anything to do with it, but also the risible notion that the oil so produced is "ethical oil" (a phrase attributed to Ezra Levant, a media gadfly notorious for his distant relationship with fact) or "responsible oil," a truly Orwellian assault on reason employed by, among others, Canada's prime minister, Stephen Harper (it is responsible because our intentions are pure and we're nice and therefore anything we produce is responsible too). Well, the tar sands might or might not be responsible, but they are big. The $180 billion or so invested by major oil companies is by far the world's largest energy project, indeed, the largest industrial project of any kind, anywhere. Its proponents suggest that it may be possible to produce four million barrels of oil a day, most of it for shipment to the United States, via rail if the pipelines don't get built, or to the coast and thence by tanker. By November 2014, the various companies were still burning through more than $15 billion a year in an effort to expand production.

For the purposes of this book, though, it is the oil sands' prodigious use of water that is the main issue.

It takes two tonnes of tar sands to produce one barrel of actual oil. To liberate the bitumen from its confining sand, the miners must blast the goop with scalding water to divide it into two streams, a small one containing heavy crude and a much larger slurry containing high concentrations of hydrocarbons, heavy metals, arsenic, selenium, and a dozen other contaminants. When the process is over, it is not really water anymore but a chemical soup of poisons. So far, somewhere around a hundred million cubic metres of water a year is "processed" in this way. About 10 percent is thought to be pure enough to return to the Athabasca River, though independent studies have made it clear that "pure" is a relative term — the water was found to contain high levels of mercury, thallium, cadmium, lead, nickel, and zinc.[16] An industry report released the previous year found, unsurprisingly, that tar sands mining was innocent — that the water was just as polluted before the miners showed up.

In any case, that was the good stuff. The rest is dumped into tailings ponds, there to sit, presumably, until Armageddon. "Ponds" is an understatement; they are really lakes, already in 2012 covering an area twice the size of Manhattan, and forecast to grow steadily until mid-century. One of the dams holding the stuff in is larger than the Hoover dam. Currently, the solution of choice is to cover the ponds' surface with a metre-thick layer of fresh water, giving the rest time to settle — somewhat. After some settling occurs, the water is recycled.

The Alberta government has licensed tar miners to extract 652 million cubic metres of water from the Athabasca River annually. In 2008, the last reliable figures to hand, the industry was using 184.3 million cubic metres, less than a third of the allotment, though still big — about a tenth of what the entire North American oil industry uses. Even so, there may not be enough overall. Mike Hightower, an engineer at Sandia National Laboratories and the reigning expert on the nexus between energy and water, told the writers for the website Circle of Blue that the limit is closer than it appears: "Canada has a lot of fresh water, but we are beginning to see limits on development of the oil sands. You will see limits where production hits a plateau and won't get above it. The point is that . . . they were talking about three million or four million barrels a day. The water resources won't allow them to go there. They will cap out at 2.5 million."[17]

But here's the other side of this debate. In March 2013, thirteen of the biggest companies operating in the Fort McMurray area, representing around 90 percent of all production, set up an unusual organization called COSIA (Canada's Oil Sands Innovation Alliance), and hired Dan Wicklum as its CEO. Wicklum is an environmental engineer (PhD in aquatic ecology), formerly with Environment Canada and Natural Resources Canada as, among other things, director of wildlife and landscape science.[18] It would be easy enough to dismiss COSIA ("Our vision is to enable responsible and sustainable development of Canada's oil sands while delivering accelerated improvement in environmental performance through collaborative action and innovation") as an industry front, a disinformation initiative, but it turns out that this is more than just a propaganda arm of the industry. Refreshingly, COSIA's charter declared up front: "Our industry has environmental impacts, which we will work to minimize." Unusually, and significantly, the member companies have waived intellectual property rights in a wide variety of inventions in order to share them with their competitors — more than five hundred initiatives that cumulatively cost somewhere around $900 million to develop. And they are putting up serious money, about half a billion, to minimize environmental impacts in four areas of focus: land remediation, water reuse and reduction, tailings, and greenhouse gases. So far, 185 separate projects have been set in train, all aimed at reducing impacts on the landscape.

And progress *is* being made. The intention is to reduce water use by 50 percent by 2022, with the baseline year being 2012. This will still be a lot of water, but it will reduce the water needed to produce one barrel of bitumen from 0.4 barrels of water today to 0.2 barrels. As well, a very large percentage of the water used is already recycled, somewhere from 80 to 85 percent in the close-to-the-surface open-pit mining, and close to 95 percent in the "in situ" operations that in this context means drilling to deeper levels. The declared goal is to recycle 100 percent of all water used. As far as the Athabasca River is concerned, the in situ sector uses no water from the river. The mining sector does — lots of it — but still only about 3.6 percent of the river's lowest flow, and about 0.1 percent at high-flow times.

On the greenhouse gases front, the declared aim is to reduce emissions below those produced by conventional oil extractions.

The oil sands/tar sands are never going to be pretty, but then nor is conventional mining. Less ugly is the best that COSIA can hope for. Still, from an environmental point of view, less ugly is considerably better than more ugly. Until we wean ourselves off fossil fuels, that's the best we can do.

I started this chapter quoting John Cherry quoting William Blake. So perhaps it is apt to finish with another of Blake's "Proverbs of Hell": *Expect poison*, Blake wrote, *from the standing water*.

8
Bulk Water Transfers

Humans have been diverting water since civilization began, since they stopped hunting and gathering and invented farming, enabling them to settle into stable communities. In most of the places early civilizations began, such as Egypt, Mesopotamia, and China, secure food supplies needed irrigated water, since rainfall was either erratic or insufficient. This meant containing water, then moving it to where it was needed. Dams provided containment. Canals, qanāts (which are generally gravity-fed underground channels), aqueducts, and later, pipes, did the diverting. The technology may have been rudimentary, but the thinking of the early hydrological engineers was as sophisticated as it needed to be. The Romans, of course, were the pre-eminent engineers of antiquity, and their network of canals and aqueducts, many of which are still in use, has hardly been surpassed.

We're bigger now, much more populous, and with access to energy undreamed of in antiquity, and our engineers have had greater dreams than were ever possible before. Which raises the question: We can dream big and build big, but should we?

In modern times, the three best-known large-scale, completed hydrological projects are the California canal and pipeline system, which takes water from the Sacramento and Colorado Rivers to Los Angeles and other drought-prone places; the Israeli National Water Carrier, which extracts water from Lake Kinneret (known outside Israel as the Sea of Galilee); and the late Libyan strongman Gadhafi's

Great Man-Made River, a series of gigantic pipelines that take fossil water from the Saharan aquifers to a new agricultural zone on the coast, a transport now fatally disrupted by the chaotic politics of Libya.

Modern California could hardly exist without bulk water transfers. The State Water Project and its 650-kilometre California Aqueduct is the largest engineering endeavour the world has yet seen, though it may be outdone, and soon, by the Chinese and the Indians. Every year, California shifts fifty-three trillion litres of water southward, capturing it behind twelve hundred dams on every river and stream of consequence before pumping it hundreds of kilometres, lifting it over some mountain ranges and under others, fitting the barren landscape with a massive caul of pipes, ditches, canals, and siphons that irrigates an agriculture that provides more than half of America's nuts, fruits, and vegetables. Not incidentally, it also waters the lawns and powers the car washes of the Los Angeles metropolis, all this in a region that receives less than forty centimetres of rain a year, when it gets any at all. About half the water Los Angeles uses comes through the aqueduct from the overstressed and over-allocated Colorado; the other half, notoriously, comes from the 375-kilometre Los Angeles Aqueduct, which siphoned water from Owens Lake and destroyed picturesque Mono Lake, turning it into a saline wasteland. Still, if the Owens Valley was destroyed, it was not via rape but through consensual sex, as a hydrologist at University of California, San Diego put it: the local residents were happy enough at the time to sell out to whoever would buy.[1]

Nor could Israel exist in its present form without its National Water Carrier, which turned fifty years old in 2014. It is nowhere near the size of California's water transfers, but it shifts around four hundred million cubic metres a year from Lake Kinneret to the cities of the coast and to the Negev, further south, through 134 kilometres of pipes, tunnels, reservoirs, and ditches. Among its consequences have been dropping water levels in the Kinneret itself, almost reaching crisis levels, and the ruination of the Jordan River and the Dead Sea. The decision a few decades ago to use the carrier to supply drinking water to the highly populated coastal cities made it a strategic asset, and Israeli security agencies have become very guarded about its capacity and reach. The construction of a series of desalination plants along the coast, now supplying almost all the country's drinking water, has lessened the strain somewhat.

Libya's Great Man-Made River was built to tap into the Saharan fossil aquifers and to take that water from deep in the desert to the coast, where new agricultural zones would be created — despite knowing full well that in thirty years, or fifty, the water would inevitably run out. Stage one, completed before Gadhafi's overthrow, was a nineteen-hundred-kilometre waterway consisting of pipes large enough to drive a truck through, carrying two million tonnes of water to the coast every day. More than four-fifths of the water would be used to irrigate new farming zones, some of them in the desert itself. The cost was somewhere around $32 billion — but then, in Colonel Gadhafi's day, oil revenues seemed even more endless than the Saharan water.

The two largest works-still-in-progress are India's vast array of pipes and canals tying virtually all of the subcontinent's rivers together into one massive interconnected network, and China's startling notion to hydrologically re-engineer the entire country, moving massive quantities of water from the more or less sodden south to the increasingly arid north. The most ambitious not-yet-built system is the Red-Dead Canal, carrying sea water from the Gulf of Aqaba to the Dead Sea, and desalinating it along the way.

These six aren't the only inter-basin water transfers that have been built or contemplated. Plenty of other pipelines thread across most continents, ranging from the functional to the bizarre. There is a new pipe from the Austrian Alps to southern Europe. Turkey's former president, Halil Turgut Özal, once proposed what he called the Peace Pipeline, which would carry more than two billion cubic metres of water a year from the Ceyhan and Seyhan Rivers and essentially spread it through the Middle East and the Gulf. The cost, before the whole thing was shelved for its political improbability, was going to be around $20 billion. As previously mentioned in Chapter 6, the Congolese dictator Mobutu once proposed a pipeline carrying Congo water through Angola, where it was hardly needed, to Namibia, Botswana, and South Africa, where it was. So far, no one has proposed emptying the Guarani aquifer and piping its water to whoever needs it, mostly because, so far, no one *does* need it.

In addition, there are many ideas that Peter Gleick has called zombie projects, unlikely and usually grandiose projects that should never be built but stubbornly refuse to die. One of the most outlandish is to shut off San Francisco Bay and turn it into a freshwater reservoir, but there are plenty of others almost as grotesque: the notion of an undersea pipeline from the Amazon to North Africa; damming the Strait of Gibraltar to dry up the Mediterranean and "create" thousands of square kilometres of new land. The list includes turning north-flowing Siberian rivers (the Ob and the Lena) southward to the Volga basin or the Aral Sea; the crazed notion of turning James Bay into a freshwater reservoir and diverting its water southward into the Great Lakes; and the NAWAPA (North America Water and Power Alliance), which takes pride of place in Gleick's chapter on zombies, taking water from British Columbia to pretty well everywhere, including California, the Mississippi, even the Great Lakes (again!).[2]

If California's water transfer system is huge, what China is proposing is more massive yet. This is the $80 billion project to re-engineer the country's fresh water, taking vast quantities from the lake-filled and often flooded south to the arid north, where much of the population lives and most of the agriculture is sited, and which has been rapidly running out of water. The North-South Water Diversion Project will subsume existing canals, among them the Grand Canal between Beijing and Hangzhou (finished around the year 500 CE). It will also require more than 3,000 kilometres of new canals, tunnels, and conduits, some of them climbing high elevations to cross the Himalayan Plateau, and in many cases having to dive beneath the rivers they must cross. The first, easternmost stage was completed in 2013, pumping fifteen billion cubic metres along 1,160 kilometres of canals, among them the Grand Canal. Most of the water pushed through the new system is highly polluted, even toxic, and must be cleaned before use. The midstream link, delayed by several years, saw its first water reach Beijing in December 2014, through another 1,300 kilometres of canals. The third (most ambitious and much-delayed westernmost) link, is the one supposed to climb to the Himalayan Plateau. Total water movement will be around forty-five billion cubic metres a year — a large number but a mere 7 percent of China's supply, nowhere near enough to solve North China's water problems. Perversely, although the capital is stressed for water (and its aging pipelines beginning to rot), water remains ridiculously cheap to consumers at less than a dollar per cubic metre.

Few outside China believe the project makes any sense, and it has many critics within China too. On the other hand, most of the politburo members are engineers, and the former president Hu Jintao, who championed the project, was a water engineer, which might explain China's predilection for massive projects. Critics point to the ecological senselessness of pushing water to the north to grow food that is then shipped back to the south; they point to the bizarre notion that the Chinese government is planning half a dozen "new cities" of more than a million each, some of them in the Gobi desert; they point to the idiocy of a policy in which water is still cheap in areas where supplies are tight; they point to the ecological hazards of mass water transfers between two utterly different natural ecosystems; and they note that the Yangtze, from which most of the water will come, is polluted to a dangerous degree.[3]

Simultaneously, the Chinese are building a seawall along the coast. When it is finished, it will cover 60 percent of the total length of the country's coastline and will be longer than the Great Wall. Their rationale is that the coastal region covers only about 13 percent of the country's landmass but produces 60 percent of the GDP — and in this ecologically primitive view, all those coastal wetlands are just taking up space. The portions already constructed have caused a dramatic decline in internationally shared biodiversity and associated ecosystem services. Virtually all environmentalists and ecologists think the seawall a bad idea, but it is proceeding nonetheless.[4]

Not to be outdone in grandiosity, India is building a pipeline from the Tehri dam in the Himalayas to divert water to the upper reaches of the Ganges River and thus supply Delhi with drinking water. This is just a small part of a much larger and more ambitious scheme, which is to build a fifteen-thousand-kilometre network of tunnels and canals to shift 174 billion cubic metres of water a year from areas with surplus water, mostly in the north, to arid regions, mostly in the west and south, such as Karnataka and Tamil Nadu. There will be, altogether, thirty new links between Himalayan-fed rivers and rivers in other basins, and the whole thing will cost, when it is finished, somewhere around $170 billion. The imperative that is driving the project is growth: the need for economic growth to feed the ever-expanding population. The Indian government calculates that it needs to increase its irrigated agriculture from around 100 million hectares to 135 million in a few decades. The National River Linking Project, as it is called, will also help to minimize flooding in the monsoon season and add thirty-four gigawatts of hydro power to the grid.[5]

Another grand dream, this time in the Middle East, is the Red-Dead Conveyance, and its sometime variant, the Med-Dead Canal. If built as originally conceived, it would be a huge desalination plant that would replenish (or fatally alter) the Dead Sea and provide emission-free electricity to both Jordan and Israel. It was originally proposed by the American engineer Walter Loudermilk in the 1950s and has been received in both the Israeli and Jordanian capitals with enthusiasm that waxed and waned over the decades according to the politics of the time. Finally, in October 2013, Jordan green-lighted the project and two months later signed a water-sharing pact with Israel and the Palestinian Authority, prompted no doubt by a raft of studies suggesting that regional temperatures would sharply increase by century's end and that rainfall could diminish by 30 percent or more.

But because this is the Holy Land and a place with a deep human history, the provenance of the project goes back a lot further than the modern era — at least, it does if you interpret scripture in a certain way. An American engineer, Randolph Gonce, a soil and water conservation specialist, became involved in plans to replenish the Dead Sea through an odd combination of familiarity with a Tennessee underground pumped-storage facility near Chattanooga called Raccoon Mountain, an interest in Chunnel engineering, news reports on the rapidly decreasing levels of the Dead Sea, and an interest in Holy Land history. He knew, for example, that King Hezekiah had built tunnels in the soft sandstone under Jerusalem to protect the Gihon Spring in the Kidron Valley, the city's water supply, and he read of the Prophet Ezekiel's curiously detailed description of a river flowing in a channel from beneath the Temple. Ezekiel measured the channel's drop at one in one thousand, and then recounted how the Dead Sea would thereafter spring to life. Gonce was taken with the fact that a tunnel bored from the Mediterranean coast to the Dead Sea would plausibly have the same slope as Ezekiel's one in one thousand, about the same drop as decent drainage for sodden agricultural fields in the modern era.

The Med-Dead tunnel, which Gonce fancifully called the Ezekiel Project, was briefly favoured by the Israelis over the Red Sea counterpart, partly because it would be entirely within state borders and it would be shorter, therefore cheaper. It would, in fact, be gravity-fed and self-sustaining after construction, unlike the Red-Dead variant. But the involvement of Jordan, the interest of international funding agencies like the World Bank, and the enticing prospect of rare political cooperation favoured the longer and more difficult version.[6]

The plan was to push a channel from the Gulf of Aqaba on the Red Sea to the Dead Sea, following more or less the Jordanian-Israeli border. To get there, it would have to be pumped up a full 200 metres, somewhere around Mount Seir, before beginning a 660-metre descent to the Dead Sea, widely known in the tourist brochures as the "lowest place on earth." The conveyor would carry just short of a billion cubic metres of water. As it dropped, gravity would generate enough power to desalinate about 40 percent of its flow, the resulting saline brine flowing into the Dead Sea and lifting it back to levels last seen in the 1960s. The idea was that the power generated and the water produced would make the project economically self-sustaining.

Naturally, critics abound. The most vocal opponent is an NGO called Friends of the Earth Middle East, based in Amman. It protests that the impact of mixing Red Sea water into the Dead Sea is, at best, unknown, though some studies have shown that the newly invigorated Dead Sea would be prone to algae blooms and gypsum crystals with uncertain but probably negative consequences.

The deal actually signed at World Bank headquarters in 2014 is rather different and much less ambitious, though it still involves dumping ocean water into the Dead Sea. It involves a conventional desalination plant in Jordan, at Aqaba, producing about forty-eight million cubic metres of fresh water for both Israel and southern Jordan; the resulting brine would then be pumped 150 kilometres to the Dead Sea, again with a view to replenishing its water (or again, contaminating it, depending on your point of view). The agreement has two other political advantages. In return for fresh water from Aqaba, Israel agreed to provide Amman, the Jordanian capital, with somewhere around thirty million cubic metres of fresh water from Lake Kinneret, and the Palestinians would be entitled to buy an allocation at "preferential prices," a phrase left awkwardly unspecified. Private companies would build the Aqaba plant at their own expense, recouping their investment through sale of the resulting fresh water. The brine pipeline would be financed by donor countries with a World Bank bridge loan. It would be located on the Jordan side of the border, thus lessening the possibility of environmentalist disruptions, as the Israelis cynically admit, Israel being more susceptible to environmentalist pressure than Jordan. Indeed, Friends of the Earth Middle East has already expressed reservations about the whole thing, and the tourist operators on the Dead Sea itself are grumpy about any change whatever.

Sometimes bulk water diversions are not physically possible or would so be so expensive that they simply price themselves out of possibility. A recent Alaskan governor, Walter Hickel, had the idea to take water from Alaska's Copper River in a pipeline down the coast past parts of western Canada, Washington, and Oregon to southern California. When the estimated cost surged past $30 billion or so, the concept lost steam — desalination, though it was then still expensive, was cheaper than *that*.

Plenty of beguiling ideas have been put forward to overcome this "not possible" problem. Why not direct glacial melt to the Atacama Desert of Chile? Ship Alaska water in supertankers to San Diego? Use giant plastic bags to tow water? If you look at the literature, it sometimes seems that shifting large quantities of water from water-rich to water-starved places is the perpetual-motion machine of hydrology, attracting all kinds of dreamers, deluded entrepreneurs, and scam artists, my favourite being the Canadian dentist who secured rights to an Iceland glacier before the securities commission caught up with him and fined him a million bucks for fraud. Speaking of water running uselessly into the sea — the notion that icebergs calved off the Greenland glaciers slowly melt fresh water into the sea as they drift southward past Newfoundland has excited the avarice of more than one would-be entrepreneur into envisaging what *Modern Farmer* magazine sardonically called "the cold rush."[7]

Why not lasso these wandering water farms and either chip them into usable chunks or tow them to someplace their meltwater would be useful? A Saudi prince, Mohammed al-Faisal, came up with this notion in the 1970s; it was he who figured he could take all those useless Antarctic icebergs and tow them up to Mecca. The study he sponsored proved conclusively that no iceberg would ever cross the equator — by the time it got past the tropics there wouldn't be enough ice left to fill a martini glass, never mind a cistern. Even to get icebergs to South Africa, which at least is in the same hemisphere, would be a challenge: a four-month journey under some of the most extreme weather conditions on earth, and to what purpose? South Africa is water stressed, but not critically so. Namibia could use the water, if it could get up the coast another thousand kilometres or so, but who in Namibia could pay for it?

Northern Hemisphere to Northern Hemisphere towing might be feasible, but which places outside the tropics needs water that badly? Simulations have been run, for example, to see if towing icebergs from

Newfoundland to the Canary Islands is feasible — but again, to what purpose and for what market? And how to deal with these behemoths when they reach their destination? Icebergs are *big*. They'd have to be mined and cut up into usable chunks, an effort that would probably cost somewhere around $10 a tonne, far more expensive than desalinating sea water. Moreover, icebergs are damnably unstable, likely to tip at any moment, putting lives at risk. And if you moored an iceberg in San Diego, to take one possible destination, you'd have to purify the water afterward because the berg would be contaminated by coastal pollution. Still, a Newfoundland company is already selling $100,000 "hair nets" that oil companies can use to tow icebergs away from drilling rigs, so the technology is feasible.

A second proposed method of getting estuary water to someplace useful would be to use superannuated oil tankers for transport. Hundreds of single-hulled tankers are being broken up in India's scrapyards or are simply rotting away in situ, so why not refit them to transport water instead of oil? Single-hulled vessels, banned for oil because of the risk of spills, would not be a problem for water — a breach would simply dump fresh water into saltwater, something that would puzzle a few fish but would otherwise cause no harm. But the cost of cleaning up each ship for water transport proved a major problem, possibly costing as much as $6 million per ship, and it has never been proven to be completely successful, at whatever price. Benzene is not something you want in your drinking water.

Some attempts have been made to use non-oil tankers. A decade or so ago, Turkey struck a deal to take surplus Turkish water to Tel Aviv in tankers and built an extravagant water-fuelling depot at Manavgat to fill them. The trade never happened — Libya's then strongman, Gadhafi, vigorously objected to *anyone* supplying Israel, and the resulting outcry killed its already remote chance of success. And in the dire Spanish drought a few years after the turn of the millennium, Barcelona was running so low on water that it hired tankers to bring water in from Marseilles. Only one made the trip. It dumped nineteen million litres of nice fresh water into the city's distribution system — and it was all used up in less than an hour, proving to everyone's satisfaction that tankering in water was a fool's game. (The city was finally rescued by a break in the drought.)

Another idea, seemingly arrived at by a number of people more or less independently at more or less the same time, was huge plastic bags for liquid transport. Sometime in the 1950s — the exact date is unclear — a Cambridge scientist originated the idea and it was picked up by the Dunlop Rubber Company of Bristol that went on to make what it called the Dracone bag, after the Greek word for sea serpent. Dracone bags were tough, long, and narrow, and could be pulled through the water at a good rate of fifteen kilometres an hour, though at that speed they thrashed about a bit (hence the name). Some were actually deployed, mostly for military transport of fuel oil and to help clean oil spills. It wasn't a big segue to try to do the same thing with fresh water, though at that time there were few accessible places that seemed to need the water that badly. The only real test of the technology to date has been the use of smaller (thirty-five thousand cubic metre) bags that for years have been chugging between Turkey and its arid province of Northern Cyprus. The operator, Nordic Water Supply, is planning to expand its operations to take water from Turkey to the United Arab Emirates.

One of the newer big-bag dreamers is businessman Terry Spragg, who earlier had joined al-Faisal to experiment with towing icebergs. He, along with everyone else who tried it, abandoned that, but water transport remained a passion. Lately, he has been trying to float, if that's the right term, a company that would employ massive plastic bags to take water from places of plenty to places of none. His Spragg Bags are modular and can be zipped together in a train, and would be towed to their destination by tug. As a test, he wanted to fill a bag or two at Turkey's Manavgat terminal and tow them down to Gaza, a notion simple enough in theory but surrounded by a thicket of prickly politics. As I write this, no test has been made.

Another outsize personality who has gravitated into the water bag business is Ric Davidge, usually described as "Alaska's former water czar," which he was, or as a "water mogul," which he is trying to be. Once a combat medic in Vietnam, he has formidable political connections, having been an adviser to Ronald Reagan and to former Alaska governor Walter Hickel. After reinventing himself several times, Davidge ended up as president of an Alaska-headquartered consortium called WorldWater, whose partners included the shipping company Nippon Yusen Kaisha and the Abdul Latif Jameel group of Jeddah, Saudi Arabia. The company's declared aim is to take water from Alaska, or wherever it is plentiful, to wherever it is needed, anywhere in the world. In any case, WorldWater has, as of 2014, yet to ship a single drop.

Of all the doers and dreamers, though, the most plausible idea remains that put forward by former oil-patch engineer James Cran, whose Medusa Corporation is headquartered in Calgary, Alberta. He has been working on the notion since the mid-1980s, when he conceived the idea of transporting water from the Columbia River estuary to San Francisco and Los Angeles. The Columbia, as he points out, is eight times the size of the Colorado, which remains the only out-of-state water supply for California, and even extracting six billion cubic metres a year would comprise only 3.7 percent of the Columbia's flow, not enough to affect either fishermen or fish. At the time, he took out a patent for his version of the giant bag, but nothing came of it.

Cran's Medusa bags — named after the jellyfish — would be massive if they ever get built, as much as 500 metres long and 150 wide, with a draft of 22 metres — they'd be as long as six football fields and carry somewhere more than a million cubic metres of water. They would be made of high-tensile polyester fibre; Cran calls them pillow tanks. They would carry fresh water at very slow speeds, somewhere around five and a half kilometres an hour, and because of their deep draft would have to be filled and emptied at offshore buoys. Medusa bags would stay afloat without help because fresh water is less dense than salt. Cran's analysis is that they would perform best when 40 percent filled, at which point they would be largely impervious to wave action — a saltwater wave could cause an internal freshwater wave, which would then cause another saltwater wave on the other side in the same direction.

Medusa has already deployed a small 4,500-tonne bag in tests off Vancouver and is planning a larger test with a 300,000-cubic-metre bag equipped with strain gauges and other devices for metering performance. The California Department of Water Resources told Cran that the state's annual ongoing water deficit, currently handled by drawing down groundwater, is around five billion cubic metres per year; he seems confident that Medusa bags could shift some 600,000 cubic metres a year by 2020, with a target of about 6 billion by 2030. More than 600 million cubic metres would require 25.5-million-tonne bags, or multiple trips by smaller numbers — substantial, but far from impossible. The source? Still the Columbia River. Whether he will get the chance remains to be seen.[8]

All of the major water transfer projects described at the start of this chapter were built without a thought for the ecological consequences — there really weren't thought to *be* any ecological consequences. Certainly none of the people who conceived and constructed them, California included, balked in any way at moving water out of one water basin and into another. We needed these massive diversions, so we built them; it is hard, now, to make a real case that we are worse off.

This doesn't mean we should do more of it. Or does it?

For better or worse, ecologists have only recently taken to examining the matter, and it is still imperfectly understood.

A fundamental principle about water management today, widely accepted by engineers and environmentalists, is that to manage global water supplies properly, we should treat the river basin as the core hydrological unit — river-basin integrity should be the first priority in supply-and-demand management. This recognition is really not political; many river basins cross international frontiers, and while there might be quarrels over allocations, the underlying principle is understood: for long-term health of the system, withdrawals must not endanger the ecological services the river provides to all riparians.

This applies not just to surface water. Groundwater is equally affected and should be treated similarly. In fact, this notion of water-basin management has been enshrined in international water law, though not always honoured.

Managed in this way, water retains at least a simulacrum of natural balance. After all, in most water uses, for agriculture or even for municipal water and wastewater supply and treatment, the water is not actually *used up*, but put to some use and then returned to the hydrological cycle, not always in pristine condition but still there — and treatable. Exporting bulk water destroys this natural balance. So again, it is widely though not universally accepted that moving a billion or so cubic metres from, say, the upper or middle reaches of a river and fluming it across a divide into another, drier, basin has large and mostly negative consequences. Not just for those creatures downstream, now deprived of the water they had been used to, but also for the river and the basin itself. It can even be argued that such export has negative consequences for the receiving basin too, in that it encourages insecure development in places where it shouldn't happen — Las Vegas, a thirsty city in a desert, is everyone's favourite example, but Los Angeles and much of the irrigated agriculture of the Central Valley are others.

To come back to the example of the American Great Lakes, eyed enviously by thirsty places in many states and viewed with anxiety by nationalist protectionists across the border in Canada. As we saw in Chapter 3, the Great Lakes do indeed hold a great stock of fresh water, almost a quarter of the global available supply, but little of it is renewable: taking a tiny few percentage points more from the inflow will eventually cause the lakes to shrink, so it is the inflow that counts, not the volume of water stored. This is poorly understood, even by professional water managers who should know better. A pristine example is the "water czar" of Las Vegas, Patricia Mulroy, who to her credit has persuaded (and bullied) the city and its casinos into becoming admirably water thrifty. Still, when she was told that states bordering the Great Lakes (and not just on the Canadian side either) were balking at devising ways to send her the water she wanted, she exploded: "We take gold, we take uranium, we take natural gas from Texas to the rest of the country. We move oil from Alaska to Mexico. But they say, 'I will not give you one drop of water!' They've got 14 percent of the population of the United States, and 20 percent of the fresh water of the world — and no one can use it but them? 'I might not need it. But I'm not sharing it!' When did it become their water anyway? It's nuts."[9] But it is not nuts. It is prudent and sensible. What is nuts is the city of Vegas itself, but that's another story.

Considerably less nuts is when water is diverted from the upper or middle reaches of a water basin to lower down in the same basin. Manhattan's water grid is an instructive example. Manhattan's water delivery tunnels, begun in 1917, were only (almost) completed in 2013. This is mass water movement in spades. New York City's nineteen reservoirs and three controlled lakes hold 2.2 trillion litres of water, spreading over eight New York State counties and stretching 150 kilometres up to the Catskill watershed. The whole thing is entirely gravity-fed, and the water can take three months to get to its final reservoir, Hillview in Yonkers, which itself holds only about a day's supply. From that last reservoir, the water flows into underground aqueducts and tunnels, the last of which, prosaically known as Tunnel #3, will not be finally completed until 2021, though in 2013 the water it carried finally reached Manhattan. Until then, the city's water supply was vulnerable, either to natural disasters or sabotage. Up to 2013, the five boroughs relied on Tunnels #1 and #2 for their drinking water, put into service in 1917 and 1936 respectively. Tunnel #3, at a construction cost of over five billion, finally provides New York with the critical redundancy it needs, and allows for repairs to be made on the other two — for the first time since they were constructed.

What if you plan to divert water from one basin to another, not from the upper reaches of a watercourse but at the point it empties into the sea? Do the usual strictures still apply? If you take, say, water from Alaska's Copper River just as it empties into Prince William Sound, does that still cause harm? Or take water from the Columbia River's estuary to San Diego down the coast, as James Cran is proposing?

You can hear many a would-be exporter complaining about all that water running "uselessly" into the sea. The Alaskan town of Sitka, which has been trying unsuccessfully to export water from nearby Blue Lake for decades, is typical, offering all and sundry more than thirty-four billion litres a year and a cost of one cent per. True, Blue Lake is not the estuary exactly, but it is not far off, and Garry White, boss of Sitka's economic development association, has argued to all comers that removing 8 percent of the watershed flow every year will not harm anything "because much of it is already being lost to the ocean" less than a mile away.[10]

But is it really "lost" to the ocean?

In some ways, the political and emotional arguments against bulk water exports are stronger than the ecological ones, often stronger than the facts warrant. It is not always easy to separate opinion from fact, to separate a dispassionate discussion of whether more water should be moved from the idea of water grabs by foreign entities taking "our" water without our permission, or under the sanction of a trade treaty like NAFTA. When I first started to write about water more than a decade ago, several companies were preparing to export the water of this or that lake (one in Ontario, up near Sault Ste. Marie, as I remember, another in Newfoundland) and sell it to thirsty customers. But it turns out that they really didn't have any customers for tankers full of water, and in any case, the economics made no sense: water is heavy — the definition of a metric tonne, after all, is one cubic metre of water — but it is not yet very valuable, and whatever profits were envisaged would be wiped out well before the tanker got anywhere. Nevertheless, the issue caused a minor uproar in the Canadian Parliament, mostly over ill-founded fears that if we allow one company to commodify and sell some of our water, much larger companies with much larger budgets will come and take much larger quantities of our water away to, presumably, somewhere in the United States. Essentially, that's what the Canadian fear came down to: that "they," the Americans, would come and take "our" water from us. Of course, if the Americans really wanted to do something like that, they could dip into "their" side of the Great Lakes, since the national frontier runs pretty much through the middle of most of the lakes. What is stopping them is the existence of the International Joint Commission, a bilateral management body always trotted out, justifiably, as one of the most successful water-management institutions anywhere. What is stopping them also is that they don't actually want to do it: US states that border the lakes are just as adamant about stopping massive extractions as any Canadian provinces are. In any case, under NAFTA, water — except for bottled water — is exempt from reciprocal trade rules, so no one can sue us for not letting them have our water. We'll come back to this issue in Chapter 12.

The ecological arguments against bulk transport of estuary water are a good deal more elusive than is the notion of abstracting water upstream. One such argument is that altering outflows damages estuary ecosystems and destroys wildlife and fish habitat. This is self-evidently true. But, as I've mentioned before, using fish as a reason not to do something that would otherwise benefit a large number of people has run out of steam as an engager of public opinion and now threatens to become merely old-fashioned, as attention is turned to more urgent matters like global warming. A second argument is a little more persuasive: that such transfers permit and even encourage us to live beyond our natural means, make us eternally dependent on energy-intensive transfer industries, and allow us to ignore the virtues of conservation. A third argument is that, if done on a large enough scale in a large enough number of rivers, the ocean and its currents might be affected, with unknowable ecological and climatic consequences. This is the same worry that plagues the project to dam the rapids near the mouth of the Congo — the freshwater plume that travels a hundred kilometres out to sea is beneficial, or so it is supposed, but beneficial to what, and to what degree?

On that, we have only speculation and worry, not data. As with so many other aspects of water management, dire need, and even perceived need, trumps ecological caution pretty much every time.

Part Three

Complications

9
Minimizing Farming's
Water Needs

Growing food for humans to eat takes enormous amounts of water — amounts so enormous, in fact, that agriculture accounts for no less than 70 percent of all human uses of water, and even more in developing countries — up to 95 percent in some cases. The average person can get by on a little more than two litres a day for drinking, but it will take somewhere between two thousand and five thousand litres to grow the food that person must eat (the higher number is mostly for meat eaters). That's about one litre of water per calorie consumed. David Molden, in a monograph for the International Water Management Institute, provides this striking image:

> Imagine a canal 10 meters deep, 100 meters wide, and 7.1 million kilometers long — long enough to encircle the globe 180 times. That is the amount of water it takes each year to produce food for today's 6.5 billion people. Add 2–3 billion people and accommodate their changing diets from cereals to more meat and vegetables and that could add another 5 million kilometers to the channel of water needed to feed the world's people.[1]

By extension, enormous numbers of people — and we're getting more numerous all the time — will multiply this number commensurately. By conservative estimates, it will take an additional 1.3 billion litres of water just to feed the extra mouths that will be born between now and 2025. That, as Sandra Postel, head of the Global Water Policy Project, once calculated in an equally striking image, is the equivalent annual flow of ninety-seven Colorado rivers — *and where are we to find ninety-seven new Colorados?* Ninety percent of the increased demand will be in developing countries.

It is these numbers that have persuaded economists to stop measuring farm yields in tonnes per hectare; water is now the limiting factor in food production, not land, and so yields are now commonly calculated in kilograms of food produced per tonne of water.

More recently, Postel has neatly reconceptualized agriculture's water problem:

> The search for solutions needs to begin with a reframing of the question. Instead of asking where we can find 97 Colorado Rivers' worth of water, the question is: *How do we provide healthy diets for 8 billion people without going deeper into water debt?* Framed this way, the solutions focus on getting more nutritional value per drop of water used in agriculture, which is the key to solving the water-food dilemma. There are many ways we can grow more food for the world with less water, with most falling into four broad categories: (1) Irrigate more efficiently; (2) boost yields on existing farms, especially rain-fed lands; (3) choose healthy, less water-intensive diets; and (4) use trade to make the smartest use of local water.[2]

I'd add a fifth: develop new salt-tolerant, water-thrifty, and high-yielding crops. And possibly a sixth: turn over management of irrigation water to those who actually use it.

Postel has made another striking calculation: Reducing irrigation water consumption by a mere 10 percent could free up enough water worldwide to meet all the new urban and industrial demands anticipated for 2025.

And 10 percent is an easy target. We can almost certainly do better.

Irrigating More Efficiently

About four-fifths of the world's croplands depend entirely on rain for water. It's the last fifth that relies on irrigation, but irrigated lands produce more than 40 percent of the world's food. Improving irrigation productivity is a good place to start looking for reform.

Until recently, irrigation techniques hadn't really changed since ancient times — and irrigation is almost as old as farming itself, going back at least five thousand years in central Asia and a great deal further in Mesopotamia. It was used in Imperial China, in old Laos, in ancient Africa, in Tanzania and Zimbabwe, and in America before the Spanish ever laid eyes on the place. For millenniums, farmers have used aqueducts and ditches to take water to where it was needed, and furrows with crude sluices to divert the water to their crops, giving them increased yields and lessening their dependence on the caprices of the weather. But on any real scale, irrigation is not much more than a century old; it needed the massive dams and water diversions of the modern era to make it possible.

The numbers are revealing: two hundred years ago, total irrigated land probably amounted to as little as 6 or 7 million hectares, not much bigger than Long Island or metropolitan Los Angeles and its exurbs. By 1900, the total had jumped to some 50 million hectares, and that number nearly doubled in the half century that followed. Of the estimated 250 million hectares currently under irrigation, half was added in the last three decades, and dozens of countries, ranging from little Israel to massive China, rely on irrigation for much of their domestic food production — Pakistan up to 80 percent. Most of North America relies on rain-fed farming, but without irrigation, agriculture in on the High Plains and in California's Central Valley, one of the world's most productive vegetable and fruit resources, would hardly be possible.

In general, farm yields on irrigated land are nearly double the yields found elsewhere. Without irrigation, yields in the world's major breadbaskets — on which the feeding of the planet is dependent — would drop by almost half. But that irrigation is not without its problems. While it is true that more land is being brought under irrigation, a good deal of formerly irrigated land is now being taken out of production too. Salination of soil to a degree that inhibits farming is spreading at a rate of almost a million hectares a year. In Pakistan alone, two million hectares have been decommissioned, the soil poisoned by high salinity, and farm yields in those area are down 30 percent. Egypt is showing similar declines. In the US Imperial Valley, more land is being decommissioned than commissioned, again because it is overly salty. Salinity and the increasing concentrations of noxious substances in water is the vulnerable underbelly of the irrigation revolution, threatening huge areas of otherwise productive lands. The problem is simple to state, though not easy to fix: without great care and skillful management, large-scale irrigation causes waterlogging, depletion, and pollution of the water supply, and rising salinity in the soil. Left unchecked, these problems can eventually kill the soil altogether.

A summary sheet put out by the US Salinity Laboratory outlines the problems succinctly:

> Application of irrigation water results in the addition of soluble salts such as sodium, calcium, magnesium, potassium, sulfate, and chloride dissolved from geologic materials with which the waters have been in contact. Evaporation and transpiration (plant uptake) of irrigation water eventually cause excessive amounts of salts to accumulate in soils unless adequate leaching and drainage are provided. Excessive soil salinity reduces yields by lowering plant stand and growth rate. Also, excess sodium under conditions of low salinity and especially high pH can promote slaking of aggregates, swelling and dispersion of soil clays, degrading soil structure, and impeding water and root penetration. Some trace constituents, such as boron, are directly toxic to plants.[3]

A second problem with irrigation is water loss that compounds the salination issue. For irrigation, river water is either diverted to canals or directly taken up by pumping. Much of that water is lost in the process. As a working average, almost 60 percent of the water intended for irrigation never gets to the croplands, and a further percentage that does get there never actually gets to the plants that need it. Leaking pipes, unlined canals, evaporation from open reservoirs and canals, and poorly directed spraying cause much of the water to be wasted. Some of this wastage returns to the groundwater, so it is not entirely lost, except in the sense of adding to farmers' costs.

None of this means that irrigation has been a bad thing or that it is necessarily doomed to failure. There are many ways of increasing water efficiency in agriculture, and all of them have a beneficial effect on salinity. The most obvious way of using water more efficiently is to reduce wastage. In many places where surface water is used for irrigation, it is brought in, sometimes for long distances, via canals or conduits, and it is far from unusual for seepage losses to amount to two-thirds of the water that sets out from reservoirs. Lining canals with plastic sheeting or with concrete can eliminate most of these losses, but it is expensive to do so. When the remaining water does get to the field, it is usually distributed in furrows, which can be diverted by the simple expedient of a shovel from one furrow to the next. Even so, furrow irrigation is wasteful, losing another 10 to 15 percent of the supplied water.

Therefore, more efficient field-distribution systems would be definitely useful. These can be either the centre-pivot system so widely used in the American Great Plains or overhead sprinklers. Both of these lose some water to evaporation, especially in dry areas, and in some jurisdictions farmers are now asked to spray at night rather than during the day to offset this. These systems spread water in the interstices between rows, and so while they are better than furrows, they are still not good. The best system is drip irrigation in which water is delivered through pipes and drip emitters directly to the roots of plants. Famously thirsty and inventive Israel pioneered the technique.

It can be done expensively, as high-end California winemakers do it (with miles of buried hoses, the water measured to a fare-thee-well, allowing the judicious admixture of tiny amounts of organic fertilizer), or cheaply, with a bucket or a barrel and a hose. Either way, it can cut irrigation water use in half. According to numbers from Lester Brown of the Earth Policy Institute, Jordanian farmers who used the technique reduced water use an average 35 percent; in India, numerous studies showed gains in water productivity ranging from 46 to 280 percent. Just as California winemakers do, other farmers have begun employing what they call "water stress management," watering only when necessary and only just enough.[4]

Another Israeli invention that has similarly increased water efficiency is the use of laser technology to get fields absolutely level, minimizing wasteful runoff and pooling. In fields that are irrigated with flood or furrow systems, this alone can decrease water use by 10 percent or so and increase yields.

One of the initiatives of Sandra Postel's project, with World Bank help, is to make drip irrigation systems as cheap as possible, and to get them to as many farmers as possible. Traditional systems are relatively expensive, somewhere around $300 per hectare. Bucket and plastic-hose devices are much cheaper, though they are also much more labour-intensive, since they have to be moved from row to row every hour or so. But in countries where labour is abundant and unemployment high, the technology works well. As Lester Brown put it, "These simple systems can pay for themselves in one year. By simultaneously reducing water costs and increasing yields, they can dramatically raise incomes of smallholders."[5]

David Bainbridge, a dryland farming specialist at the University of California, Riverside, believes that drip irrigation is not always useful in remote and low-tech parts of the world — the pipes tend to clog and animals eat the tubing. He has experimented with an even lower-tech solution: the buried clay-pot system that he believes is excellent for small-scale farmers and gardeners. Water seeps out through the wall of a buried unglazed clay pot at a rate that is influenced by the plant's water needs: "When I began the buried clay pot trials, I found that the water efficiency was good, and I kept up with the research. In India, for example, melon yield with the buried clay pot system was 25 tons/hectare using only 2 centimeters of water, compared with yields of 33 tons/hectare using 26 centimeters of water with flood irrigation. But unfortunately scientists don't typically study these traditional practices." More recently, he has experimented with what he calls wick irrigation, in which a series of holes is punched into the buried pot and porous wicks inserted, providing a slow and steady source of moisture to plants. "I conducted experiments in the California desert on these systems and found 2-10 times greater efficiency than drip irrigation systems," he says.[6]

Another low-tech device that can take the guesswork out of irrigation (and thus uses less water) is something called a tensiometer, typically a sealed water-filled porous cup with a vacuum gauge at the top; as soil moisture decreases, the water level in the tube of the gauge goes down. Some of the newer and simpler tensiometers do away with the gauges and have colour-coded bands instead, making them both easier to read and less prone to fail. In a test for the Columbia Water Project, farmers who used them reported an average of 22 percent water savings and 24 percent energy savings.

Boosting Yields on Rain-Fed Lands
Can rain-fed (non-irrigated) farming improve water use in the same way? After all, 60 percent of the world's food is still produced with rain as the primary water source. Farming techniques in much of the areas that depend on rain haven't changed in millenniums — in fact, Rome at the time of Christ was slightly more productive than most farmers in poor countries today. David Molden has calculated that three-quarters of the extra food we will need over the next few decades could be produced just by bringing the productivity of low-yield farms to within 80 percent of high-yield farms. Interestingly, new US figures suggest that although massive agribusiness farms produce more calories per dollar invested, small farms using mixed

systems of crops and livestock can actually produce more calories per hectare, done, as Molden suggests, with better water management "and non-miraculous changes in policy and production techniques." These include improving soil moisture by capturing local rain in small reservoirs to apply to crops when the rains fail through some of the low-cost irrigation systems outlined above. Small but efficient farms in the United States can produce 1.3 kilograms of wheat per tonne of water used; by comparison, in the Punjab area of Pakistan and India, the yield is less than half a kilo.

Sometimes boosting yields can be as simple as managing farming techniques. A system called System of Rice Intensification (SRI) reliably produces higher than normal yields with no fertilizer and no pesticides. It depends on four interlocking ideas, all of which sound too simple but which have been amply proven in trials. First, select and nurture only healthy plants at the start of the growing cycle and transplant them at an earlier age; second, reduce plant numbers — plant single seedlings instead of clumps, giving each plant more room to grow above and below the ground; third, enrich the soil with organic matter, keeping it well aerated; finally, avoid flooding plants and apply water parsimoniously, in ways that favour root and soil-microbial growth.

A young farmer from the state of Bihar, in northern India, Sumant Kumar, came to hold the world record for crop yield on his small-holding, using SRI techniques. He managed to get 22.4 tonnes per hectare (he only has two hectares), whereas even with commercial fertilizers, average yields hardly exceed 8 tonnes — and the average yield worldwide is only 4 tonnes. SRI techniques have since been used in a variety of environments, including the arid north of Mali around Timbuktu. The system is labour-intensive and resistant to large-scale production, but its supporters point out that millions of small farms have more available labour than money, and scaling up is not necessarily in their purview.

Another way of managing farming techniques and to preserve water is to use more under-cropping, agroforestry, and cover crops, and to practise no-till farming.

A simple technique called STRIPS, which stands for, obscurely, Science-based Trials of Rowcrops Integrated with Prairie Strips, has boosted farm yields "spectacularly," according to farmers who have tried it. All it amounts to is taking around 10 percent of the planted hectarage, usually the least productive part, and replanting it with indigenous prairie plants. The *New York Times's* Mark Bittman quotes a plant researcher at Iowa State University, Lisa Schulte Moore: "It's well-known that perennials provide a broader sweep of ecological function than annuals, so our hypothesis was that if you put a little bit of perennials — a little bit of prairie — in the right place, you get these disproportionate benefits. That is, without taking much land out of production, you get a lot of environmental benefit." The numbers seem, as Bittman put it, impressive: if you convert 10 percent of a field of row crops to prairie, soil loss can be reduced by up to 95 percent, nutrient loss by 80 to 90 percent, and water runoff by 44 percent. Biodiversity nearly quadruples, and some of those species are pollinators, predators of pests, or both.[7]

Even simpler is adopting a system of no-till farming: cultivating perennial crops that need no soil cutting. Tillage eliminates weeds but also exposes non living organic matter to oxygen, releasing their carbon dioxide. David Montgomery, a scientist with the Earth and Space Sciences Department at the University of Washington, in Seattle, has compiled data from around the world that suggest pretty conclusively that conventional agriculture degrades soil much faster than it can be created. Montgomery has concluded that no-till farming could reduce erosion to levels close to soil production rates, and that organic farming methods have been shown to be capable of preserving — and in the case of degraded soils, improving — soil fertility.[8] An Australian farmer, Angus Maurice, calls this "no-kill" farming. Wes Jackson, who has worked in the field of sustainable agriculture for more than three decades (one of his creations is the Land Institute, in Salina, Kansas), is another practitioner of no-till farming, though he calls it "natural systems agriculture." His idea is to develop perennial versions of the major food crops like wheat, sorghum, corn, and rice, crops that have deep roots and don't have to be replanted each year, thus doing away with the need to till the soil while also radically reducing the need for additions of water. The ancillary benefit is obvious: eliminating nitrogen fertilizers alone would remove two gigatonnes of carbon dioxide a year from the global atmosphere.

Choosing Less Water-Intensive Diets

This is Postel's third rail, and it risks taking us into nanny-ish territory. It is certainly true that each gram of protein consumed through rice takes five times less water than a gram of beef, or, if you want to measure calories instead of pure protein, it takes twenty times more water per calorie to take beef to table than rice. Two food scientists, Wesley Wallender of University of California, Davis, and Daniel Renault, recently with the Food and Agriculture Organization (FAO), have calculated, surely in jest, that a "balanced diet" consisting entirely of groundnuts, potatoes, onions, and carrots could be produced by using about a thousand litres of water per person per day, compared with the consumption of the steak-eating Americans or Argentinians, who currently need fifty-four hundred litres per day. They suggest that if everyone in the developed world reduced eating animal products by a quarter, it would generate 22 percent of all the additional water requirements expected by 2025. (They don't explain how water saved in, say, the United States, would benefit anyone anywhere else, since there is no way of transporting that newly liberated water any place it might be needed.)[9] Postel herself, who clearly is fond of her Colorado analogy, says that if all Americans cut back on animal products by half, water demand would be reduced by 261 billion cubic metres a year, "a savings equal to the annual flow of 14 Colorado Rivers."

It must be said that this is the least probable of the reforms. The evidence suggests, in fact, that the trend is the other way, toward more meat eating, as more societies grow wealthier and adopt Western eating habits.

Using Trade to Make the Smartest Use of Water

"Virtual water" is the water embedded in a product — that is, the water used to grow grain or to raise a cow or to make an iPhone. When that product is imported by another country, the embedded water is imported with it: it becomes water the importing country didn't need to use to acquire the product concerned. This has often been interpreted as a modern form of colonialism, as rich countries once more exploit poor countries by depriving them of a precious resource. In some cases, it is: growing roses in water-scarce Kenya for export to water-rich England may be good for trade but it is horrid for water. But in most cases it is nothing of the kind; it is merely exploiting the differentials in water productivity in different places. If a water-rich country can export embedded water to a water-poor country, it becomes a solution to shortages, an indirect means of transferring water in bulk from one place to another. Israel, for example, has chosen not to grow grain, a water-hungry crop, instead importing it from a country like Canada that has no such shortage. Israel saves between one thousand and three thousand cubic metres of water for every tonne of grain imported. Sandra Postel has estimated that about a quarter of the world's grain trade is by countries choosing to import water indirectly in the form of grain. As she points out, this can be a good strategy to avoid over-pumping of groundwater or diverting rivers.

There's another advantage. Virtual water is a way of storing water from good years for use in bad. Except where irrigation is used, all farming areas have poor seasons and good ones, and food storage is a way of smoothing the differences. The numbers are not inconsequential. The FAO's Daniel Renault calculates that stored grain alone represents a "virtual reservoir" of 500 billion cubic metres of water, or 500 cubic kilometres, a number that rises to 830 billion cubic metres if you add in sugar, meat, and edible oils — this is almost 14 percent of water in existing reservoirs.[10] Virtual water is also a way of cheaply desalinating seawater. Somewhere around 8 percent of all virtual water imports is embedded in seafood — which is water that doesn't come from any country, wet or dry, but represents a net gain of usable water for drylanders.

The only real downside of virtual water is that it risks contributing to regional food insecurity. Countries that come to depend on importing food for their people are taking a chance on prices remaining stable. As countries like Pakistan and China that have large populations and are also water stressed turn to the international grain market, prices will go up. Add this to increasing transportation costs and poor countries risk being shut out from the food market. As Postel put it in her monograph for the Post Carbon Institute, "The food riots that erupted in Haiti, Senegal, Mauritania, and some half dozen other countries as grain prices climbed in 2007 and 2008 are likely a harbinger of what is to come and suggest that a degree of food self-sufficiency may be crucial to food security."[11] Things have eased somewhat since 2009. But in 2014, the world's food experts were predicting another sharp rise in prices, and that this one would likely stay. A billion people were going hungry, not because there was too little food but because they couldn't afford to buy it.

Developing New Crops

A fifth way of saving water in agriculture is to grow crops only in those regions to which they are suited, and then, more controversially, to develop new crops suited to arid and saline conditions.

In and around Beijing, to take an obvious example, the regional authorities have banned the growing of rice, a water hog, in favour of more efficient grains. Egypt, a country with no rainfall and only one river, has also banned rice growing, relying instead on imports. In many other parts of the world, cheap and subsidized agricultural water has encouraged the growing of thirsty crops like sugar and cotton in areas without adequate rainfall, such as the American Southwest and the prairies around the Aral Sea. Removing the subsidies make these easy and obvious fixes.

The world also needs to develop new crops entirely, crops that can tolerate the more hostile conditions expected by climatologists or can colonize areas more conventional plants cannot. This is not so much a management issue as it is a science issue, albeit science snarled in politics.

Consider the Green Revolution of the 1970s that ushered in the era of relative abundance with hyper-productive grain varietals, first tested in Mexico and then used, with spectacular success, in Asia. It's true that the Green Revolution also depended on profligate use of chemical fertilizers and insecticides for its success, but the core was the new grain varieties, developed by food scientist Norman Borlaug, who won the Nobel Prize for his efforts. The Green Revolution saved millions of lives — India and Pakistan particularly would have faced catastrophe without it. It bought us decades of time in which to solve the underlying problems faced by global food production.

Now, we are on the brink of a new agricultural revolution, often called the Blue Revolution.

To some degree, this extends Borlaug's work, and is a more sophisticated version of what farmers have been doing for millenniums, fine-tuning crops through trial and selection and retrial. An example is the work of scientist Richard Richards, a plant biologist at the Commonwealth Scientific and Industrial Research Organization in Australia. Richards has bred a wheat grain with longer than normal seed sprouts, or coleoptiles, that allow the plant to reach deeper into the soil earlier than conventional seeds. In a series of trials, the new grain increased yields up to 20 percent in arid conditions.[12]

Another useful example is from Texel, in north Holland, where entrepreneur Marc van Rijsselberghe, together with a plant biologist from Amsterdam's Free University, Arjen de Vos, has developed a salt-tolerant potato. In 2014, several tonnes of the new potato were on their way to Pakistan to plant in a plot abandoned because of high salinity. There was nothing genetically modified about this new crop, just age-old breeder's techniques, inspired, as de Vos put it, by studying sea cabbages.[13]

New and better crops will also come from genetic-modification techniques that aim to introduce beneficial traits from other species and by doing so to increase yields and drought resistance, and boost nutritional values. As I wrote in another context, in my book *Our Way Out*, any results forthcoming will be controversial:

> Few fields are as rife with emotions, some of them bitterly antagonistic and even violent, as GM [genetically modified] foods. Scientists who develop GM crops will bring down a global network of anti-biotech activists on their heads; their labs will be picketed, their field trials sabotaged and their email inboxes flooded with hate mail, actions driven by a toxic mix of emotions, including fear and paranoia, and by "facts" that have only a distant relationship with truth. As a result, many of the GM labs have been obliged to surround themselves with a massive security apparatus, and operate in a world of finely-tuned suspicion.[14]

These scientists' position has not been helped by the arrogance, and propensity to lie, of the big agricultural chemical companies that are their patrons. Worse, companies like Monsanto are allowed to patent seed-grain varieties, keeping tight control of their use, and they willingly prosecute farmers into whose fields one of their patented products strays, however innocently. This patenting of life forms has locked farmers into buying annual seeds from the company, and has prevented them replicating their own field grains. To their opponents, such cycles are *prima facie* evidence of the degeneration of farming into profit. It is all made worse by the often exultant air of some company pronouncements, so when Monsanto's Robert Fraley is quoted as saying, "We're now poised to see probably the greatest period of fundamental scientific advance in farming in the history of agriculture," his opponents hear only the profit-mills grinding.[15]

It is not hard to find GM successes, even from such as Monsanto. For example, company researchers found a way to mitigate the effect of drought on corn, and in field trials, their new variety maintained yields under harsh test conditions. They did this by engineering the corn to express a protein from the bacterium *Bacillus subtilis*, an extra gene that boosted yields by 6 to 10 percent. Monsanto even got permission from the Mexican government to plant test plots in a range of northern states — a significant concession, since GM maize had been banned in Mexico for more than a decade.

Early in 2015, India, formerly a staunch GM opponent, reversed policy under its newish prime minister, Narendra Modi. Genetically modified mustard was planted at the Indian Agricultural Research Institute, and was expected to be released for sale by mid-year. "Field trials are already on because our mandate is to find out a scientific review, a scientific evaluation," Environment Minister Prakash Javadekar told Reuters late in February. "Confined, safe field trials are on. It's a long process to find out whether it is fully safe or not." Allowing GM crops is critical to Modi's goal of boosting dismal farm productivity in India, where urbanization is devouring arable land. When Modi was chief minister of Gujarat State, he had already allowed GM cotton to be planted there, and it was, according to state officials, a smashing success.[16]

Sound policy should separate the two streams of genetic modification, first by refusing to grant patents for any attempts to create varieties that are merely resistant to one or other proprietary pesticide, or that confer inbuilt commercial advantage, or that seek to control life forms. However, we should also recognize that genetic manipulation can be beneficial, and approve attempts to create transgenic varieties that increase yields, confer resistance to viral diseases, and thrive in poor farming conditions — and from which farmers can derive their own seeds.

Make Irrigation-System Responsibility a Local Affair

The last way of improving water productivity is a simple management change: move the responsibility for managing irrigation systems from government agencies to local water users' associations. This suggestion, from the head of the Earth Policy Institute, Lester Brown, has been amply proven in many jurisdictions:

Since local people have an economic stake in good water management, they typically do a better job than a distant government agency. In some countries, membership includes representatives of municipal governments and other users in addition to farmers. Mexico is a leader in this movement. As of 2002, more than 80 percent of Mexico's publicly irrigated land was managed by farmers' associations. One advantage of this shift for the government is that the cost of maintaining the irrigation system is assumed locally, reducing the drain on the treasury. This also means that associations need to charge more for irrigation water. Even so, for farmers the advantages of managing their water supply more than outweigh this additional expenditure.[17]

In sharp contrast to ecologist Garrett Hardin's tragedy of the commons theory, when local farmers assume responsibility for the water table, they generally don't rush to deplete it. Instead, they tend to conserve it, using less water overall.

People worry about agriculture more than any other aspect of the water crisis. How will we find the water to grow the food we all need? After all, droughts have killed millions for centuries — and there are more of us all the time, and more droughts too. But with careful management and sufficient resources, the food problem is tractable. Even in arid parts of the world.

10
Climate Change

In the spring of 2014, an acrimonious debate erupted in the scientific literature and the blogosphere over the causes of the enduring drought in the American Southwest, then into its third year. No one was disputing the severity of the drought, though the one in the 1930s was more severe and the drought of the 1990s more prolonged, at least so far. Towns all over a clutch of western states had to ration water. California for the first time was invoking its right to stop the depletion of underground water sources. Governor Jerry Brown declared a state of emergency. President Obama came carrying aid. Lake Mead, the Colorado's major reservoir and the source of drinking water for a dozen cities, including the ever-expanding Las Vegas, had sunk to levels not experienced since the Hoover dam was built, threatening even its hydroelectric capabilities.

A few sturdy libertarian fruitcakes aside, watering lawns in daytime became an anti-social act.[1] A winemaker of my acquaintance was obliged to thin his 2014 crop almost to nothing just to keep the vines alive, and even then he had to tanker in water when his wells started to sputter. Oprah Winfrey's decision to do the same tanker-thing for her mansion's swimming pool brought down unaccustomed negative publicity on her head. In farm country, spray-irrigation-system salespeople were falling on hard times, but the drip irrigation guys were going gangbusters. A long-hoped-for El Niño failed to materialize — or was late, or was weak — late in the year it was still uncertain, except that the rains never came, except in the occasional, and rather useless, torrent. Car washes started to recycle water. Washing sidewalks became a criminal offence.

Everyone agreed the drought was awful, but they couldn't agree on what caused it. Was it natural variability, some hitherto unexplored climatic cycle, or was it climate change? If the latter, then it raised the unnerving possibility that it might *never go away*. No wonder emotions were running high. The precipitating event of the climate-change dust-up was a posting on the White House website, a six-page "rebuttal" to a University of Colorado professor of environmental studies, Roger Pielke, self-described as a long-time analyst of climate-related disaster losses. Pielke had testified before a congressional committee the previous year and his three main conclusions were these: first, there exists exceedingly little scientific support for claims that hurricanes, tornadoes, floods, and drought have increased in frequency or intensity on climate timescales either in the United States or globally; second, on climate timescales, it is incorrect to link the increasing costs of disasters with the emission of greenhouse gases; and third, that "these conclusions are supported by a broad scientific consensus, including that recently reported by the Intergovernmental Panel on Climate Change (IPCC) in its fifth assessment report (2013) as well as in its recent special report on extreme events (2012)." The first two assertions went by largely without comment; they were similar enough to that of other skeptics summoned by the same congressional committee that they raised few eyebrows. The third conclusion, though: Had the IPCC "supported" these conclusions, really? John Holdren, Obama's chief science adviser, had said before the same committee (rather gently, I thought) that Pielke was "outside the scientific mainstream." Pielke, rather less restrained, accused Holdren of indulging in voodoo science. At this point, *New York Times* environmental reporter Andrew Revkin weighed in but made the elementary mistake of respectfully quoting Pielke, bringing down some of the calumny on his own head, the angriest from a thinkprogress.org blogger named Joe Romm, who dismissed Pielke as a mere "political scientist" (a low blow that!) and called Revkin a naïf.

Some of the disagreements over drought go a little deeper than this, though. It was interesting that two peer-reviewed papers published in 2012 came to almost opposite conclusions about global droughts. A paper in *Nature* carried the headline, "Little Change in Global Drought over the Past 60 Years"; another paper, in the journal *Nature Climate Change*, was titled "Increasing Drought under Global Warming in Observations and Models." Both papers were cited in a review subsequently published in *Nature Climate Change*, neutrally subtitled "Global Warming and Changes in Drought," which in turn came to a third conclusion, that "increased heating from global warming may not cause droughts but it is expected that when droughts occur they are apt to set in quicker and be more intense."[2]

Revkin, manfully resisting slamming Romm for calling him a Pielke toady, had another go at the issue, this time invoking two more scientists, both with reputations rather hard to denigrate. The first was Martin Hoerling, who studies climate extremes for NOAA, the National Oceanic and Atmospheric Administration, which also runs the National Hurricane Center; the other was Richard Seager, a climate scientist who studies water issues at the Lamont-Doherty Earth Observatory of Columbia University. Seager had earlier told another *New York Times* reporter, Justin Gillis, that he was "pretty sure" the severity of the current drought is due to natural variability rather than climate change.[3]

Nevertheless, both scientists agreed with the *Nature Climate Change* paper that climate change has exacerbating effects. Water is being used more intensively by a larger population than in earlier droughts, reducing resiliency. Further, higher temperatures evaporate water faster, melt snow faster, and dry out the ground faster.[4]

So far, I've been concentrating on the American Southwest. But, in all this, California and its neighbours are just stand-ins for droughts elsewhere, some of them just as prolonged and sometimes even more severe: Australia, of course — the Millennium Drought, the formerly Fertile Crescent (Iraq, Syria, parts of Turkey), southwestern Brazil, much of southern India, parts of the Sahel, the North China plains, southern Africa — many droughts, in all regions. Not all of them are caused by climate change and a warming globe, but they all exist in in an increasingly warmer environment. Which, yes, makes them worse. Possibly much worse.

So climate change is not off the hook. Some years ago, the hard-headed folk of the insurance industry, led by Munich Re, the world's largest reinsurer, got together with a clutch of climate, weather, and atmospheric scientists to see if they could arrive at a consensus about the damage climate change might be doing. The meeting concluded with a carefully worded consensus, that "changing patterns of extreme weather events are drivers for recent increases in global losses." Climate change, the meeting suggested, may not be the dominant factor, "[but] it has become clear that a relevant portion of damages can be attributed to global warming." Munich Re should know — the company has been diligently accumulating catastrophe data over millenniums. (It has a database dating back to the year 79, listing twenty-two thousand natural disasters.) As Quirin Schiermeier reported in *Nature*, the insurance giant's NatCat service shows that the frequency of weather-related catastrophes has increased sixfold since the 1950s, while the number of non-weather-related incidents (volcanoes, earthquakes, and the rest) increased only marginally.[5]

What are these "weather-related catastrophes"? Storms — hurricanes and typhoons, also tornadoes, and in winter, ice storms and severe blizzards. Exceptional rains leading to flooding, mudslides, and dam breaches. Storm surges damaging coastal communities (and, in the much longer term, rising sea levels drowning whole regions or even countries). Drought, of course.

Another set of hard heads can be found at the World Bank. In an analysis co-authored by Jamal Saghir, chair of the bank's water sector board, the bank acknowledges the obvious, that climate change is real, and suggests that "taking prudent measures to plan for and adapt to climate change must become an integral part of the Bank's water practice." And the report adds:

There is now ample evidence that increased hydrologic variability and change in climate has and will continue have a profound impact on the water sector through the hydrologic cycle, water availability, water demand, and water allocation at the global, regional, basin, and local levels. Many economies are at risk of significant episodic shocks and worsened chronic water scarcity and security. This can have direct and severe ramifications on the economy, poverty, public health and ecosystem viability.[6]

At the same time, the report also recognizes that climate was not the only risk, though it was an enabler of risk:

Future water availability and use will also depend on non-climatic factors. Climate change is only one of many factors that will determine future patterns of water availability and use. In the absence of policy changes, non-climatic factors are likely to aggravate or attenuate the adverse effects of climate change on water availability and quality, as well as have a significant influence on water demand. Population growth and economic development will play a dominant role. Non-climatic impacts could be generated through many realms — from population growth, migration and income to technologies and infrastructure to land-use patterns and agricultural activities/irrigation. Such non-climatic drivers could dwarf the impacts attributed to climate change alone.

So what *is* the IPCC's current assessment of these issues?

Floods and Droughts
Elementary physics would suggest that as the globe warms, evaporation will increase. No one disputes that, but measuring or modelling the increases are much trickier, and so is figuring out where

the extra water will go. It is known that the moisture content of the lower atmosphere has been increasing at about 1 percent per decade, at least since the 1980s, and this is true for the air over land as well as over the oceans. The likelihood is that for every degree the planet warms, total evaporation, and therefore precipitation, will increase by about 2 percent. Much of that, of course, will fall back on the sea, simply because the sea covers more area than land. But where it will fall on land is harder to say. Climate modelling is still not fine-grained enough to make accurate regional projections, but some kind of consensus has emerged: higher latitudes, from about 30° north to about 85° north, things are going to get wetter, and in lower latitudes, from about 10° south to 30° north, things are going to get drier, perhaps a lot drier. The Southern Hemisphere is not quite a mirror image — ocean currents and landforms are quite different — but the same tendencies are likely there too. Those two trends, more water in higher latitudes, less in lower latitudes, have actually been intensifying since about 1900.

There is also considerable evidence that the heavier rain in higher latitudes, as well as the scantier rain in lower latitudes, will come in what the IPCC calls "heavy or extreme precipitation events" — that is, in monsoon-style downpours. The expectation is that such events will increase about 7 percent for every degree the planet warms. There is certainly plenty of anecdotal evidence that this is happening. In July 2014, parts of Connecticut and Long Island got more than a dozen inches of rain in a few hours, and California's welcome but insufficient downpour in December was similarly intense. The consequences are easy enough to see: flash floods, roads and bridges washed away, crops eroded and possibly destroyed. Many regions may have to become dependent for their water on a few massive storms instead of on steady rain. England, to take an example from recent years, will have to become more adept at flood management.

These heavier downpours may not be all bad. An intriguing study in *Nature Climate Change* suggests that torrential rainfall is more efficient at restoring depleting aquifers than had been assumed, better even than steadier precipitation. This goes against the conventional wisdom, which has it that the heavier rains ran off the land into the oceans before they could affect groundwater. The study was done over a fifty-five-year period in Tanzania, in what the authors call the "semi-arid tropics," where "episodic recharge interrupts multiannual recessions in groundwater levels, maintaining the water security of the groundwater-dependent communities in this region." Their conclusion was also counterintuitive: increased use of aquifers would likely be a viable adaptation to increasing variability and potential shortages.[7]

The flipside of flooding is, of course, drought. As the heated debate over the California drought shows, it can be difficult to pinpoint real causes. The measure most commonly used is called the Palmer Drought Severity Index (PDSI), named after the hydrologist Wayne Palmer, who invented it. But it is not very easy to apply, being actually three indexes in one: the Palmer-Z Index, which tries to show how short-term (monthly) moisture conditions depart from normal; the Palmer Drought Index, which show longer-term (annual) changes in moisture conditions; and the Palmer Hydrological Drought Index, which reflects groundwater conditions, reservoir levels, and more. The validity of the Palmer results depends on accuracy of data input, a scientific truism after all, and from most parts of the world data are spotty, relying too much on anecdotal evidence. Still, the IPCC has concluded that the PDSI for the planet as a whole shows droughts increasing in length and severity, though their confidence level for this assumption is not very high — "the patterns are complex." As with precipitation, some areas have become wetter, especially in higher latitudes, while others have become notably drier. Globally, soil moisture has decreased. Interestingly, droughts in the United States, except for in the Southwest, have decreased.

Storms
One of the most common assumptions is that tropical cyclones are increasing in numbers and severity. This assumption is wrong in its essence, though not in all its details.

First of all, there is actually no clear trend in the frequency of tropical cyclones, according to the IPCC. Some years there are plenty, in other years not. In 2005 there were so many that the National Hurricane Center ran out of names and had to start the alphabet again; in 2013, there were only a handful. In 2014, the eastern Pacific reached Y in the alphabet, the Atlantic only H. There is some evidence, though it is disputed, that the number of cyclones will go down as a consequence of climate change, since high-altitude winds are likely to become more prevalent, which would undercut storm development — hurricanes, powerful as they are, are nevertheless vulnerable to wind shear. It is true that the number of Atlantic hurricanes has been increasing since about the 1970s. On the other hand, that increase was predicted by long-term climatic cycles, and likely has nothing to do with global warming.

The evidence is not much better that hurricanes have become more intense as the planet warms. On the contrary, a new study in the *Bulletin of the American Meteorological Society* in November 2014 that assessed cloud-top temperature data over thirty years found "zero increase worldwide in the average potential intensity of storms."[8] In the Atlantic basin, only five Category 5 hurricanes have ever made landfall on continental America, although, of course, accurate records don't go back much before 1900. The first was an unnamed storm that struck the Florida Keys in 1935, when the barometer fell to an extraordinarily low 89,200 pascals (892 millibars). The second was Hurricane Camille in 1969, still the strongest Atlantic storm ever recorded, carrying winds gusting to 305 kilometres an hour and pushing a storm surge that reached 7.6 metres above mean tide levels. The other three, it is true, were in the last few decades: Hurricane Andrew of 1992, and two in 2007, Hurricane Dean, which hit the Yucatan with winds of 296 kilometres an hour, and a week later, Hurricane Felix, which struck as far south as Nicaragua and Honduras, a most unusual track.

Nevertheless, a few things are happening as a result of climate change that impact on the severity of storms. The first is warming sea surface temperatures, or SSTs. A precondition for hurricane formation is that the ocean surface should be at least twenty-six degrees Celsius, and preferably twenty-six and a half or higher. No one really knows why this number is the magic one. It has to do with the climatological factors governing tropical oceans. Temperatures can be higher than twenty-six, but not lower — the higher they are, the greater the potential for damaging convection currents to occur. Higher temperatures don't increase the probability that a system will coalesce into a hurricane, but they will tend to make that hurricane more intense. If the seas warm further, warmer water will be found further north, which means that cyclone formation can then occur outside what has historically been the "normal" tracks. Worse, at least for those of us living on Atlantic shores in latitudes higher than, say, 40 North, is that hurricanes approaching our latitudes will no longer cross waters cool enough to reduce their intensity. That is, more strong hurricanes will reach further north than before.[9]

The second consequence of climate change is that the lower troposphere carries more moisture than before, and moisture is the fuel for hurricanes.

On tornadoes, the evidence is much clearer. They have not gotten bigger or more frequent. There are actually fewer days with tornadoes than there were, but more days with clusters of tornadoes, rather than singletons. No one knows why this is so.

Glaciers and Snow Cover

On the topic of glaciers and snow cover, the IPCC is confident: continuing decrease in glacial mass — but not everywhere. Decrease in snow cover — but not in all regions. Earlier peak runoff from glaciers and snowmelt — pretty much everywhere.

What is also clear is that glacial shrinking is already having dire consequences, and not just because glaciers are pretty to look at and will contribute to sea-level rises (almost a third of the rise to date is caused by glacial melt). Melting is consequential because of the number of people who depend on the reliable water glaciers provide through seasonal melting for their livelihoods, indeed, for their lives. This is as true of the rich world (western Canadian prairie water is at least partly dependent on glacial melt) as it is of the poor world — Nepal, Bolivia, and other Andean countries. Moreover, shrinking snow cover in multiple places is not just a question of having to move ski resorts to higher elevations; the snowpack keeps water through the wet season to provide it in the dry and hot season, and this reliability has kept farmers, towns, and cities supplied with predicable water. If it melts faster, new reservoirs will have to be built — and we know the animus against new dams.

There are thought to be somewhere around 150,000 glaciers in the world. Except the high-altitude Himalayan glaciers, above fifty-five hundred metres, pretty much all of them are shrinking. Some disappearing glaciers and snowcaps are more visible than others. Tanzania's Kilimanjaro is perhaps the most obvious because it is visible from afar to the thousands of tourists who come for the surrounding game parks. The Ruwenzori mountain glaciers on the border between Uganda and Congo are melting just as fast, but hardly anyone from the outside world notices because tourism there is minuscule. Glaciers at polar latitudes and at exceptional elevations shrink slowly, whereas the world's tropical glaciers (71 percent of them located in Peru) are disappearing more rapidly. Some, such as Chacaltaya in Bolivia, have disappeared altogether, leaving what was once the world's highest altitude ski area still high but now dry, "the lodge, still stocked with rental gear and decorated with ski murals, sits mostly abandoned" — this from an excellent report by Elisabeth Rosenthal, in the *New York Times*.[10]

The state of the Andean glaciers is dire. The World Bank, not usually given to alarmist prose, calls the disappearance of South American glaciers a "slow-moving catastrophe."[11] During heavy rain, as the bank pointed out, the ice masses on the mountains store the precipitation as snow that melts in the dry season and feeds the water into the rivers below. When the glaciers no longer do this — and they are forecast to disappear completely within thirty years — they will no longer provide the main source of water for around forty million people from Columbia to Ecuador, Peru, Bolivia, and Chile.

El Alto, Bolivia, the poorer suburb of La Paz, could become the first major population centre to become a casualty of climate change.[12] The damage could be mitigated by prudent management and the construction of reservoirs, but there is scant sign of that happening.

Peru, a well-managed country with ample resources, is doing better. Three of the country's watersheds have set up programs to work with local people to adapt to more efficient irrigation and better conservation. One of the aims is to experiment with crops that use less water, and to reforest where practical. Regionally, Bolivia, Ecuador, and Peru have set up a joint study project on glacial retreat with financing from the Japanese government and the Global Environment Fund.

Mitigation is the aim, not prevention. It is too late for that.

The issue in the high Himalayas is quite different and commonly misunderstood. Richard Armstrong, a glaciologist at the CIRES National Snow and Ice Data Center who has studied the matter, found many misconceptions. (CIRES, or Cooperative Institute for Research in Environmental Studies, is a joint project of the National Oceanic and Atmospheric Administration and the University of Colorado, by no coincidence the home university of Roger Pielke of drought fame.) "We've all heard these stories about how Himalayan glaciers are melting faster than anywhere else, drinking water is disappearing, there will be widespread catastrophic floods, etc.," Armstrong said. "Well, when you start looking, there's really no data to support those statements."[13]

Glaciers above fifty-five hundred metres are not melting and may even be growing in mass. On the other hand, more than half the glacial ice at those altitudes never melts, at any time of the year, making its stored water more or less useless for humans. In fact, glacial melt contributes only a few percentage points of stream flow in the lower river basins. As the globe warms, more of the precipitation lower down will fall as rain and not as snow, and that rain will have to be captured and stored if it is to be used. This means more dams, and quickly.

European glaciers, for their part, are in full retreat — more than half the ice has vanished since comprehensive mapping began in the late nineteenth century. The Norwegians produce most of their electricity from glacial-melt hydro stations and are contemplating big increases in wind power as that resource diminishes. In Switzerland and Austria, mountain and ski resorts are attempting to adapt, by moving higher up the slopes, by creating more snow of their own, or, in a couple of particularly pathetic cases, by laying insulating blankets over the ski runs to prevent melting. The Italian ski resort at Vedretta Piana is using snowplows to push snow off the glacier higher up the slopes — hastening the demise of the glacier itself.[14]

North American glaciers have been retreating an average of eighteen metres a year since about 1950, when careful measurement began. Once, there were 150 ice sheets in Glacier National Park, straddling the border between British Columbia and Washington State. When measured a decade or so ago, there were 37. In 2014, there were 25, and in 30 years there may be none. All the park's glaciers have retreated dramatically since the middle of the nineteenth century. Canada's Athabasca Glacier, an outflow from the Columbia Icefield in Jasper National Park, is the most visited (and the most measured) glacier on the continent. Because more snow still falls in a year than can melt in the short summers, the ice still accumulates at high altitudes and creeps forward at a couple of centimetres a day, spilling (if that's the right word for such stately movement) from the icefield over three bedrock steps, a very slow and cold waterfall. Despite all this, the tongue of the glacier is melting fast — the glacier has lost half its volume and has retreated one and a half kilometres over the last 125 years. Every year, British Columbia's glaciers shed the equivalent of 10 percent of the Mississippi River's flow because of melting.

Despite a brutally cold winter in 2013-2014 (caused by an unusual polar vortex, itself plausibly caused by melting Arctic ice), the snow on North America's mountains continued its retreat. In the last forty-seven years, in the calculation of Porter Fox of *Powder* magazine, the skiing bible, two and a half million square kilometres of spring snow cover has disappeared in the Northern Hemisphere. It is likely, Fox says, that somewhere between 25 and 100 percent of snow at America's ski resorts will disappear by 2100.

As Fox wrote in an op-ed in the *New York Times*,

> I was floored by how much snow had already disappeared from the planet, not to mention how much was predicted to melt in my lifetime. The ski season in parts of British Columbia is four to five weeks shorter than it was 50 years ago, and in eastern Canada, the season is predicted to drop to less than two months by mid-century. At Lake Tahoe, spring now arrives two and a half weeks earlier, and some computer models predict that the Pacific Northwest will receive 40 to 70 percent less snow by 2050. If greenhouse gas emissions continue to rise — they grew 41 percent between 1990 and 2008 — then snowfall, winter and skiing will no longer exist as we know them by the end of the century.[15]

The median date of snowmelt in the Colorado Rockies shifted two to three weeks earlier from 1978 to 2007. In Washington, the Cascades lost nearly a quarter of their snowpack from 1930 to 2007.[16]

Should the water world be concerned that ski resorts might close and that sales of snowmobiles are going down the toilet? Yes. Snowmelt in the Sierras (and other ranges along the west coast of America) feed the Colorado and other rivers, and contributes precious water to one of the most productive agricultural zones on earth. Less snow means less food — or more expensive food, as other, more difficult, water sources are found or created. This is true in the Caucasus, the Alps, the Nepalese Himalayas, and elsewhere. There are ways to mitigate the damage — more storage is one of them — but they will be expensive and of uncertain utility.

Sea-Level Rise

The "debate" about sea-level rise has been so extravagant on both sides that it would be entertaining if the consequences weren't so serious. In a way, Al Gore started it. His suggestion that sea levels would rise six metres "in the near future," swamping London and a baker's dozen other cities (and putting New York skyscrapers half under water, according to one *An Inconvenient Truth* graphic) was widely panned by reputable scientists, and drew a flood of retaliatory condemnation down on his head, a flood whose eddies were still being felt in 2014. Christopher Booker, writing in the reliably conservative *Daily Telegraph* in London, called sea-level rise "the greatest lie ever told."[17] As whoppers go, this statement must itself be right up there with the best of them. The *Telegraph*'s equally reliable junk-science-loving companion, the weekly *Spectator* magazine, ran a cover story featuring the same anti-hero, "the Swedish geologist and physicist Nils-Axel Mörner, formerly chairman of the International Union for Quaternary Science's (INQUA) Commission on Sea Level Change," who was able to assure readers that this whole notion of sea-level rise was "nothing but a colossal scare story" and that sea levels hadn't changed in fifty years. Never mind that INQUA has disavowed Mörner, has disbanded its Commission on Sea Level Change, and has felt the need to strongly affirm its belief in both climate change and rising sea levels — none of this made it through Booker's ideological filters.

So what is the truth?

The truth is, the truth is rather hard to measure. The sea is a restless thing, with strong and weak currents, tides that depend on moon phases, an overturning circulation that periodically brings deep cold water to the surface (such as, El Niño), and other perturbations. The North Atlantic gyre means, for example, that the centre of the ocean is higher than its perimeter. All this is complicated by the fact that land is rising in spots after being crushed under the last ice age. Or sinking, like Nova Scotia, which has finished its rebound and is now settling back down again. All of this is complicated further by the fact that certain parts of the oceans are saltier than others, and this makes measuring pan-ocean densities very difficult.

Nevertheless, metadata from dozens of studies show more or less the same thing: the ocean levels are rising, though not to Gore-ish levels. Much of the rise to date, about a third of a metre per century, is due to thermal expansion as the oceans warm, and to glacial melt.

Not much, so far, has been due to Greenland and Antarctic melt, but when that happens, the rise will likely accelerate. The IPCC's estimate is that the rise will reach two-thirds of a metre per century by 2100, which would double the current rate, "but the upper end could be much higher." These numbers were confirmed early in 2015 by a team of scientists from Harvard, with one particularly scary confirmation: the rise *is already* accelerating.[18]

The IPCC's report lays out a number of scenarios. On the Pollyanna-ish side, the world gets its emissions under control, limiting the increase to twenty-five centimetres this century, a little more than the twenty centimetres of the twentieth century, but not catastrophically so. On the gloomier side, if emissions continue unchecked, sea levels could rise up to ninety centimetres this century. That, *pace* Al Gore, would put many of the world's major cities at risk, including London, Shanghai, Venice, Sydney, Miami, New Orleans, and New York. Entire countries like Bangladesh would be underwater.

Even the half metre we've had to date has been causing coastal erosion and increased damage because of storm surges. For the provision of fresh water to people who currently have little or have unsafe supplies, the worst consequence will be the diversion of precious funds and energy to solve a problem not of their making.

Global warming may not cause droughts, but it does make them worse. It may not cause more storms, but it is extending their range. It does cause glacial melt and reduced snow cover, which will complicate almost any proposed solutions for water security. It makes flooding worse, which is a problem even in developed parts of the world: the devastating flood in Calgary, Alberta, in 2013 flooded the homes of the rich more than of the poor — the rich had chosen the "desirable" locations along the river.[19] It does cause sea levels to rise, and that is going to make everything worse. Perhaps some places will benefit. Perhaps we will be able to grow pinot noir grapes in northern Canada, fresh ginger in Nova Scotia. But it seems idiotic to count on it.

11
More People, Using Ever More Water

More people means more water will be needed for drinking and for sanitation. More people need more water to grow more food. More people means more industry, and that means more water to make stuff, and then to clean up the residues that making stuff makes. This much is obvious.

But how much more water? If it takes somewhere between two thousand and five thousand litres of water to produce a single person's daily food, can we just multiply that figure by nine or ten billion, representing the people who will be living on the planet — and eating food — by 2015, to get a global need? Can we deduce how much water they will need, agriculture aside? Can we figure out how many trillions of litres we all will need and where it will be found?

This is difficult because there are so many variables and regional disparities.

We don't *really* know how many people will be alive in 2050. We don't *really* know how much water they will need per capita. We don't know if the current disparities in consumption — food aside, hundreds of litres a day per person in the rich world, a few dozen at most elsewhere — will persist. We don't really know how much the idea of water thrift will take hold, or by how much we can cut overall consumption. We don't know how much water industry will use, or other non-agricultural practices. We don't *really* know what the diet of the nine billion will be like — if the billions of poor become poor no more and start eating a Western diet, water use will skyrocket. (As Andrew Revkin once put it on his blog, "Nine billion vegan monks would have a far different greenhouse-gas imprint than a similar number of people living high on the hog.")[1]

And we don't really know what the supply will be either. How much (extra) water will come from the skies in the form of rain or snow? Will there be more usable water than now? Or less, because droughts will consume productive farmland? And we don't really know how much water agriculture will use, except to say, as I did in Chapter 9, that it will be more than now but less per capita than now. Farming is already making improvements in efficiency and will no doubt make more, but how much will be enough?

Water is more plentiful in some parts of the world than in others, obviously enough. Some of those wetter parts have greater than average population densities, some far smaller. Some wetter parts conserve water well. Some drier parts, perhaps out of necessity, do the same. There are profligate users everywhere, and thrifty ones, in wet places as well as dry. You can't shift water from thrifty places to profligate ones — that would rather spoil the point of the thriftiness, even if shifting the water were plausible.

Still, it remains a truism that, globally, more people will need more water, maybe as much or more as Sandra Postel's "20 new Nile Rivers or 97 new Colorados," which she suggested would be needed just to grow the food for the new mouths to be born in the next few decades. So it seems useful to ask the question, how many more people, really? And where will they be found?

When I wrote about population in an earlier book, my purpose was to see what the effects of exponentially growing populations would have on the world economy and the global ecology — I was arguing for a post-growth social order. So in that book I raised three sets of questions:

The first had merely to do with numbers: How accurate are our population numbers, and how reliable are the projections?

Then: How many is too many? How many is too many *in the light of the other crises that face us*? How many is enough? What is our responsibility to the other species that share our planet? How do we manage raw numbers and consumption levels?

The third set was more complicated: How do we get there? Can whatever number we consider correct be achieved without coercive policies? Is coercion to be contemplated? What policies are available? How do you deal with population momentum (continuing growth even after fertility rates drop), with demographic shifts caused by population policies, and with aging populations caused by lower fertility rates? What will happen to social mobility in a static population were we to achieve it? Will population reduction, or stasis, happen all by itself? Could we actually cope with dropping population levels?

It is beyond the prescription of this book to suggest strategies for reducing population, or even to ask whether such a reduction is desirable. So here, only the first set of those questions matters: What are our numbers and what are the best guesses going forward?[2] And a second set I didn't ask in the earlier book: What is a reasonable minimum target for per capita consumption that will still yield a comfortable way of life?

So, How Many Are We?

On September 11, 2014, at 16.58.57 UTC, the day and time I began this paragraph, the population of the planet was 7,260,437,120, a nicely precise, if fictional, number that would be larger by eight or nine or ten a few seconds later, and larger by more than 108,000 or so just a day later. Globally, we added slightly more than eighty million in 2013, about the population of the United Kingdom, with the Republic of Ireland thrown in. We're expected to add somewhere around two billion more by 2050, rounding it out to around nine and a quarter billion. Or maybe more.

It took all of human history until 1830 for world population to reach one billion. To reach the second took one hundred years; the third, thirty years; the fourth, fifteen years; the fifth, a mere twelve years. The "explosion" in human population coincides with the invention of the first mechanical steam engine, the real start of the Anthropocene Era.

More humans will be born in the next thirty-five years than were alive on the planet in 1900.

A stable population is theoretically achieved with a birth rate of 2.1 children per woman. In 2008, the UN was projecting that the rate would decline to 1.85 per woman by 2050. In 2009, it upped the projection to 2.01, still below replacement, then a year later raised it again to 2.12, and a year after that it was back down again. In 2014, it was up sharply as new numbers came in. All this variability is based on observed trends in a few countries, some of which were experiencing birth rates far below replacement (the countries of the former Soviet Union, Germany, Japan, and a few other developed countries), while others, such as Niger and Kenya, were still up there at better than four.

There were also some anomalous trends. For example, fertility in most developed countries had been declining for several decades, but then, oddly, started to climb again. The favoured explanation was that they had dropped as more women entered the workforce but rose again as a consequence of a benign public policy providing good-quality universal daycare. In 2008, for example, UK figures showed that Britain's fertility rate had climbed back up to just under two births per woman, the highest it had been since the 1970s — and the net population increase was the highest since 1962.

To see what a difference small changes in fertility make, consider Kenya. Instead of declining as expected, Kenyan fertility actually rose between 1998 and 2003, from 4.7 to 4.8, an apparently tiny increase. But because the UN had assumed a reduction, the forecast population number for Kenya in 2050 had to be increased dramatically, from forty-four million in a 2002 projection to eighty-three million two years later. In neighbouring Uganda, where the president, Yoweri Museveni, had been calling for higher fertility to produce the people he believes are necessary to fuel economic growth, the trends are even more drastic. The UN's projections assumed that Uganda's fertility rate would decline by 60 percent by 2050, bringing the country's rate to somewhere around three children per woman. But between 1960 and 2003, the decline was a miserable 3 percent — Uganda's seven million in 1960 became twenty-nine million in 2005 and even if the UN's projected decline comes into effect, it would still triple to ninety-one million by 2050.

In 2014, the projections shifted yet again. In some studies, they varied downward. In others, upward.

Early in the year, demographers issued new numbers, suggesting that the global fertility growth rate was somewhere around 1.14 percent per year, down from its peak of 2.19 in the 1960s (note that these are percentage increases, not live births per woman). The same study suggested that, by 2020, the growth rate would be less than 1 percent and only half a percent by 2050.

Even those low numbers didn't imply precipitous population decline, at least not globally, because in some areas, mostly Africa, the rate remained stubbornly high. In some places it *will* actually decline, as it has in Japan for two decades. Western Europe will plausibly go down to less than 400 million by the end of the century, from around 460 million now. China's rate is already below replacement, at about 1.25 live births per woman. Germany's rate is as low as 1.36, Italy's 1.4, Spain's 1.48. In other parts of the world, the rate of increase will decline too, but not nearly as fast. Projections for Africa have been notoriously unreliable (and depend to a large degree on imponderables such as education levels and robust rates of economic growth), but the population of about 1.1 billion in 2014 will very likely increase to 4.2 billion by 2100 — that is, almost all of the global population growth predicted by 2100 will come from Africa.

But the degree of uncertainty is still high. The UN's 2014 figures, published in September, carried a more sombre headline: "World Population Stabilization Unlikely this Century." They were now estimating eleven billion by 2100 and possibly much more. Then, a study in *PNAS*, using a computer model based on WHO and US Census Bureau numbers, declared that "under current conditions of fertility, mortality and mother's average age at first childbirth, the world's population would be 10.4 billion by 2100." In a rather grim image, the study said this: "Even a world war that killed as many as

85 million would barely register a blip on the population trajectory this century — the losses, huge as they would be, would be obliterated in less than a year."[3]

Population momentum has something to do with this. Even with global fertility rates at just the replacement level, age momentum would carry the population to almost nine billion by 2050. To put it another way, if the population growth rate declines from 2.1 percent to 1.8 percent, the population's doubling time has been extended from thirty-three years to thirty-nine. So somewhere less or more than ten billion by 2050. What does this mean for water?

Most of the increase in population will come from sub-Saharan Africa, but also from Southeast Asia and Latin America, precisely the regions where water stress levels are highest. Does this mean that people in Europe and North America — the rich world — should stop worrying about either water or population? No. We should continue our efforts to use less and to conserve more. California, to take an obvious example, can almost certainly conserve enough water to offset shortfalls caused by the current drought. We should continue to worry, but also recognize that reducing our own consumption will not help other parts of the world.

Water stress comes about in either of two ways. The first is physical scarcity, where there really isn't enough water to go around. The other is economic water security, which is essentially poor management — there is enough water, but it is not going where it is needed, for essentially the same reasons that pollution abatement is so difficult in so many countries — lack of competency in government, but mostly a crushing poverty that can make it almost impossible to raise the capital needed to fix the problem. The more people there are, the harder it gets.

Of the forty-five most water-stressed countries, thirty-five are in Africa, most of the rest in the Middle East or central Asia, and even there, water is present in sufficient quantity, if only just. In many countries, governments give their people no water at all — they must fend (and fetch) for themselves. In many countries too, the available water is polluted and unsafe to drink; or so meagre that all they can do is drink, not wash or flush; or a supply that is so expensive that it consumes the major share of what small incomes exist. But they all get *some* water. So while water scarcity affects every continent, and a billion people in the world are under water stress (led by Ethiopia, Haiti, and Niger), no one is actually dying of thirst. Of contamination, yes; of disease, yes; but not of thirst.

Not yet.

In parts of Africa and Asia, even in parts of China, a daily allocation can be as little as twenty litres for a family, which means there is no water for washing or for sanitation, never mind for laundry and bathing. In other parts of the world, as we well know, running a shower for an hour or filling a swimming pool are tasks that are done with no thought whatever. But — to repeat — you cannot just average the numbers and get everyone enough.

The Consumption Numbers Game

Just as it is hard to give accurate population numbers, it is difficult to make accurate assumptions about how much water is actually consumed and by whom. As with population, arriving at useful data is to make a series of more or less plausible assumptions and to allow for gaps, for unknowns.

To illustrate the point, most tables of per capita water use in the world measure withdrawals from known sources. Water use for agriculture is measured, but only where farming draws from aquifers, rivers, or reservoirs — rainfall doesn't factor into the calculations. So even comparing similar countries becomes tricky. For example, the United States' per capita water use is between 1,500 and 1,630 cubic metres per person per year; Canada, about 1,330. So the United States is more "wasteful" than Canada? Perhaps. But consider that, overall, the United States is hotter than Canada and its water demand therefore commensurately greater. Consider also that Canadian agriculture is largely rain-fed — only about 12 percent of national water withdrawals are for farming, well below the world average of about 70 percent, and well below that of the United States, where farming accounts for 41 percent of water withdrawals. That doesn't make Canadian farmers better water stewards; it just means that, in the United States, farming is done in hotter and drier regions and is more dependent on irrigation. Therefore it uses more groundwater and water from rivers.

You can also look at Canada and the United States and conclude that both are profligate compared with a country like Denmark, where each citizen uses nine times less than each Canadian. Comparisons of eight countries with comparable affluence levels shows the United States using the most per capita water (425 litres per person per day), with Canada a close second (326 litres per day). Even in the United Kingdom, the average person uses only 150 litres per day.

Accurate consumption figures for poor countries hardly exist, even as guesses.

Another factor in usage calculations is that water consumption generally goes up as national affluence increases. Industrial consumption of water, for example, has been shown to increase as nations get richer — industrial uses are around 10 percent of consumption in poor countries and can reach 30 or 40 percent or even more in rich countries. But in the rich-rich countries, the percentage used by industry starts to go down again. US industry now uses around 43 percent of all water withdrawals, down a few percentage points even in the past decade, attributable to improved efficiency — modern factories need less water to operate. Overall water use goes down too as cities improve infrastructure, stop more leaks, and mandate more efficient appliances for domestic uses.

Affluence can also overcome water stress. The Gulf nations of Bahrain, Qatar, Kuwait, and Saudi Arabia are rated among the world's most water-stressed countries with the least available water per capita, but their natural shortages are overcome by importing food (virtual water imports) and by spending vast sums to produce more (desalination). When a country in the same region is not affluent, these remedies are not within reach, which is why Yemen threatens to be the first country to run out of water entirely.

Good management can also overcome stress. Singapore is densely populated, with no natural water (rivers or aquifers) of its own, but its investments in desal technology, its international supply agreements (with Malaysia, mostly), and its skillful management have enabled it to be water-secure.

The great mega-slums of the developing world are not produced by overpopulation alone; they are more a consequence of a general failure of governance, a cruel aid regime, and ignorance — collateral damage of capitalism's relentless predation, millions driven into cities by the collapse of traditional farming, and the arrival of a cash economy. Consequently, the cities are expanding faster than governments can cope, even were they willing. Similarly, while exponentially expanding population is not the only cause of global poverty, it *is* one cause. And from an available water point of view, the poor become a problem only when they try to get richer, which in all justice we should help them become.

Everyone has water — they must, or they'd be dead. Millions have only just enough. Millions of people have no safe water. Millions have only polluted water. Burgeoning populations will add more millions to the millions already facing shortfalls in safe water — more millions who will be without the most basic of basic amenities. Those extra millions won't make water problems impossible to fix. Only much harder.

Part Four

Flashpoints

12
Scarcity, Yes; Conflicts, Maybe

Water shortages cause water tensions — that much is obvious. But do water tensions lead to water conflicts or even war, and under what circumstances? If so, where are they most likely?

As I wrote in the first chapter, there are two distinct but overlapping problems with the world's water: one is dirty water, the contamination of water supplies even in areas where water itself is relatively abundant; the other is where water, dirty or clean, is in short supply. The same two crises can be causes of water tensions too, complicated by local, regional, national, and sometimes transnational tensions. Sometimes these may arise from water directly; in others, water becomes a surrogate for greater conflicts.

Small and local doesn't necessarily make the problems more tractable. In Cochabamba, the insertion of a profit-taking multinational into an area of great poverty set off rioting, a brief army occupation, and some entirely unnecessary deaths, followed by regime change. Water riots broke out in Mauritania when the ruling classes, mostly Arab, were perceived to be keeping water from the poor, mostly black, and again when desperate nomads from Mali crossed the border when their wells ran dry. Still in Mali, nomadic Fulani herdsmen were met by armed bands of Dogon pastoralists whose water they were stealing, not having any of their own. Livestock herders in Mongolia's South Gobi Desert have been conducting raids on multinational mining companies over water abstractions. (So far, the herders, equipped with cell phones and an acute sense of their own entitlement, seem to be winning.) There are dozens of episodes like it, in Yemen, Pakistan, Syria, and many others places.

Some countries that are water stressed are managing badly; others are doing well. Qatar, almost always described as "oil-rich Qatar," is a champion mismanager. It hardly ever rains in the Gulf (only eight centimetres a year, on average), the country has no lakes or rivers, and what meagre shallow aquifers once existed have long since been depleted. There is some deeper groundwater but not enough, and the country's drinking water is supplied through energy-hungry desalination plants that consume more than a fifth of the country's electricity generation. As in Saudi Arabia itself, this raises the unnerving prospect that the two will cease to be major exporters of petroleum because they will need what oil they have to produce the water without which they cannot survive. Nonetheless, in defiance of all pricing and conservation logic, water in unlimited quantities remains free to all Qataris. Singapore, on the other hand, is everyone's poster child for smart management and thrift. The American Southwest falls somewhere in the middle — shortages there have been caused by poor assumptions made over the last 150 years that growth and water supply would never collide and that the climate would remain unchanged and unmodified. On the other hand, while the Americans may have miserably dysfunctional government, they have expertise and money aplenty and are, if rather belatedly, responding with the proper sense of urgency.

River basins and aquifers that cross national frontiers are the principal international flashpoint. In the last few years, violence or threats of violence have affected many parts of the world.

The most contentious are the same handful that have been sore points for decades. The Tigris and Euphrates basins, for example, affect Turkey, Syria, and Iraq. The Nile affects six countries but mostly Ethiopia, Sudan, and Egypt — Egypt has famously declared it would go to war to safeguard its rights to the Nile that are in 2014 once again threatened by the surprising fact that peace has broken out upstream.

The Jordan basin (mostly the Jordan and Yarmuk Rivers) affects Israel, Jordan, Palestine, and Lebanon; water has already been used as a weapon for war in the region. The Ganges and Brahmaputra Rivers are a main contention between India and Bangladesh. The Indus, which rises in Tibet but flows through India and Pakistan to its delta, raises tensions not only between India and Pakistan but also internally in India between the Sikh-dominated Punjab and the Hindu Haryana (a dispute that led, among other things, to the assassination of the Indian prime minister, Indira Gandhi). In central Asia, the Amu Darya and Syr Darya, two rivers whose diversions caused the destruction of the world's fourth largest lake, the Aral Sea, are now the source of disputes between a clutch of post-Soviet independent countries. In 2012, despite the region being proclaimed a success of integrated water resources management (water-management partnerships and local committees have helped raise water productivity by 30 percent), there were serious clashes between Tajikistan, Kyrgyzstan, and Uzbekistan over water-allocation issues. Threats were delivered over cutting supplies, with retaliatory threats uttered about shutting down natural gas deliveries. On the other side of the Caspian, the Kura-Araks system, a major water source for Armenia, Georgia, and Azerbaijan, is also the subject of international wrangles. We will return to these. Elsewhere there is tension over the Okavango River, which acts as the border between Namibia and its two otherwise amiable neighbours, Botswana and Angola, and the Salween shared by China, Myanmar, and Thailand. Uganda has deployed army units to stop incursions of water-stressed herders from neighbouring Kenya.[1]

In Chapter 5, I quoted Ismail Serageldin, then a World Bank vice-president, as saying, "The wars of the 21st century will be fought over water." Well, it remains to be seen — we've got 85 percent of the century to go — but so far, as many have pointed out, shooting wars just haven't happened. A scrutiny of almost two thousand "water-related events" since 1950, in fact shows that two-thirds of all encounters were cooperative rather than antagonistic. More than 150 international water treaties were signed in the same period.

The Pacific Institute's Peter Gleick has compiled every researcher's favourite water-conflict compendium, a comprehensive chronology dating back to 3000 BCE (that was when Ea, a Noah precursor, punished humanity with a six-day rainstorm). Ever since, water has been used as a weapon (poisoned wells, diverted canals, deliberate floods, and so on).

The first real "water war" was between two Mesopotamian city states, Lagash and Umma, in which Umma's water was diverted into boundary canals to starve it out — and that was forty-five hundred years ago. But there have been many violent events around water in the interim, and, as Gleick pointed out in an email to me, they seem to be increasing. It is also true that Israel did exchange artillery and rocket fire with Syria over water, and it is now widely accepted that the 1967 Arab-Israeli war had its roots in water conflicts as well as in territorial and security concerns. It's also true that many dire warnings have been issued about the Nile River, among them by two UN secretaries-general, Boutros Boutros-Ghali and Kofi Annan, and Egypt has many times warned that attempts to tamper with its allocation of Nile water would be a *casus belli*. And it is also true that population and development pressures have made available per capita water increasingly scarce. And if two-thirds of water encounters have been amiable, by extension a third were not. Shots have been fired over water, dams have been blown up, water supplies tampered with or cut off, wells filled in or poisoned, hydroelectric stations bombed.

But water wars, in modern times?

Mark Zeitoun, a Canadian engineer-turned-social-scientist now working at the Water Security Research Centre at the University of East Anglia, has spent decades considering the notion of conflicts over water and the perennial and deeply political question of who gets how much water, and why.

Zeitoun's contribution is to analyze water disputes that are neither the much-feared water wars nor the much-lauded examples of cross-boundary cooperation. In his view, conventional analysis downplays the role of "power asymmetry" between contesting nations. "The reason these conflicts fall short of war and are largely silent may have much more to do with the imbalance of power between the riparians than with a perceived cooperation between them," as he put it.

With this in mind, he has suggested the notion of hydro-hegemony as a way of framing the discussion. As he wrote in a co-authored 2005 paper,

> Hydro-hegemony is hegemony at the river-basin level, achieved through water resource control strategies such as resource capture, integration and containment. The strategies are executed through an array of tactics (e.g., coercion/pressure, treaties, knowledge construction, etc.) that are enabled by the exploitation of existing power asymmetries within a weak international institutional context....
>
> A few key questions may illustrate the point: If Turkey — upstream on the Tigris and the Euphrates — can build the GAP (Guneydogu Anadolu Projesi, the Turkish acronym for the South-eastern Anatolia Project), what is preventing Ethiopia from doing the same on the Nile? How is it that the Palestinians living on the West Bank of the Jordan River cannot approach the river, much less pump from it? The answers are found in power play. Power relations between riparians are the prime determinants of the degree of control over water resources that each riparian attains. Riparian position and the potential to exploit the water through hydraulic infrastructures also have some influence but are not determining except insofar as they are power related. *In brief, upstreamers use water to get more power, downstreamers use power to get more water.*
>
> We emphasize that the absence of war does not mean the absence of conflict... to suggest that war is not a likely outcome of water disputes... is not to deny the passions that international water quite legitimately arouses. Virtual water

BACK TO THE WELL

and second-order resources become, in this light, useful
mitigators of conflict-induced water scarcity but do little
to address — and in some cases may even prolong — the
conflict. Furthermore, we agree with others who empha-
size the point that a significant factor preventing war over
water is that the actions of non-hegemonic states usually
comply with the order preferred by the hegemon, whose
superior power position effectively discourages any violent
resistance against the order.[2]

In other words, "peace" can be enforced by coercion, rather than by
agreement or war.

Zeitoun's point is that once you understand not only why conflicts
happen but how "resolution" is avoided or arrived at by the more
powerful countries, it becomes easier to edge toward more genuine
cooperation. That is, "somewhere between unilateralism and comprehensive accords" (a phrase from Princeton scholar John Waterbury),
modest steps can be made toward cooperation.

As Sandra Postel and Aaron Wolf pointed out in a piece for *Foreign
Policy* magazine, "Whether or not water scarcity causes outright warfare between nations in the years ahead, it already causes enough
violence and conflict within nations to threaten social and political
stability. And... today's civil conflicts have a nasty habit of spilling
over borders and becoming tomorrow's international wars. [Also]
water disputes between countries... have fueled decades of regional
tensions, thwarted economic development, and risked provoking
larger conflict."[3] A National Intelligence Estimate prepared for the
US State Department in 2012 concluded that

during the next 10 years, many countries important to the
United States will experience water problems — shortages,
poor water quality, or floods — that will risk instability
and state failure, increase regional tensions, and distract
them from working with the United States on important

US policy objectives. Between now and 2040, fresh water availability will not keep up with demand absent more effective management of water resources. Water problems will hinder the ability of key countries to produce food and generate energy, posing a risk to global food markets and hobbling economic growth. As a result of demographic and economic development pressures, North Africa, the Middle East, and South Asia will face major challenges coping with water problems.[4]

An earlier OECD report, in 2005, pointed to another negative effect of water shortages — good development averted or delayed: "Tensions over water in an international river basin often mean that a shared resource is *not* developed. They result in, and are exacerbated by, a lack of structural stability, where capable, accountable, and responsive structures exist to peacefully manage and mitigate conflict."[5]

What follows is a summary of the main global flashpoints, or "the bad bits," as Pakistan's water minister once called them.[6]

Israel and Its Neighbours

Toward the end of 2013, an Israeli ecologist and his counterpart in Gaza wrote an op-ed piece in the *New York Times* titled "Gaza Need Not Be a Sewer." But the piece was less positive than the headline: Gaza was already a sewer, caused by the usual things bedevilling the region — Israeli intransigence, Hamas pig-headedness, Palestinian Authority vacillation, poverty, burgeoning population (1.7 million now, growing at better than 3 percent a year in a territory only forty kilometres by ten, with an eleven-kilometre border with Egypt and a fifty-one-kilometre border with Israel), and overall carelessness and neglect of the fragile environment.

When the piece appeared, Gaza hadn't enough fuel to run both its electricity supply and its water and sewage facilities, mostly because of the Israeli and Egyptian blockades. Hamas, which governs Gaza, refused to contemplate buying alternate fuels because the taxes they generated would go to its political rival, the Fatah-controlled Palestinian Authority. As a result, pumping stations stopped operating in November 2013 and human excrement began flooding into the streets. I was struck by this plaintive paragraph in the *Times* piece, authored by Alon Tal and Yousef Abu-Mayla: "Aside from humanitarian decency, there are ample pragmatic reasons for Israel to be concerned. Every day, 90,000 cubic metres of sewage pours into the Mediterranean. Israel's own drinking-water supply is increasingly dependent on seawater desalination. One of its largest facilities, in Ashkelon, is just a few miles north along the coast from Gaza. Erecting a fence can prevent terrorist infiltration, but it can't stop the flow of feces."[7]

The flow of feces, yes. Drifting feces are impervious to political control.

No one in the region seems to have learned this simple ecological fact, not yet.

The Middle East has always been the place where water wars were perceived to be most probable. As pointed out above, the Six Day War had its roots in water politics. Israel controls the Golan Heights for its water as well as for reasons of military security. In fact, the boundaries of the state of Israel are to some degree the result of water considerations.

The key is the Jordan River, which begins in three headwater streams and provides about 30 percent of the region's water: the Hasbani originates in Syria and has at least a part of its outflow in Lebanon; the Dan and the Banias Rivers originate in the Golan Heights and flow into the Jordan above Lake Kinneret. The lower Jordan is fed from springs and runoff from the West Bank, Syrian, and Jordanian waters, and by the Yarmuk River, which rises in Syria, borders Jordan, Syria, and the Golan Heights (and so Israel), and empties into the Jordan at Adam Bridge. The Jordan Valley is thus an international drainage basin. This was acknowledged by the early Zionists. Chaim Weizmann, from the beginning of his quest for a Jewish national home, asserted his desire to have control of the valley of the Litani, for a distance of about 25 miles above the bend, and the western and southern slopes of Mount

Hermon. Control of the Litani, the Jordan, and the Yarmuk was seen as critical to the future state's security.[8] David Ben-Gurion, Israel's first prime minister, often demanded the same thing, and demanded too that the boundaries of Israel include the southern banks of the Litani. That one he didn't get — it remains wholly in Lebanon. Israel has, on and off, established a "security zone" in Lebanon, and access to Litani waters was certainly one reason for doing so.

The whole region is drawing down its aquifers beyond replacement rates at somewhere around 15 percent a year, with the obvious result that water is complicating already difficult politics. If your taps run once a week and Israel's taps seven days, from the same source, that's a potent grievance to nurture. And it's not just the aquifers that are being depleted: in 1953, the Jordan had an average flow of 1,250 million cubic metres at the Allenby Bridge near the Dead Sea; it now records flows fluctuating between 140 and 170 million cubic metres, an eighth of what it was. And the water in Israel's National Water Carrier system is not as pristine as it should be; it contains mineral concentrations higher than is considered safe in Europe or the United States.

Nearly three-quarters of Israeli-Palestine water is groundwater, drawn from three aquifers, the Mountain, Eastern, and Coastal aquifers.

The Mountain aquifer is by the far the biggest of the three and for decades gave Israel about a quarter of its water — rather less now, since desalination plants are providing most of the country's drinking water. For more than a decade, water engineers have warned that the Mountain aquifer has been overdrawn to the extent of permanently imperilling its existence. Worse, municipally-added chlorine and ill-treated sewage are both finding their way into the aquifer, as are traces of nitrates, soluble organic material, and heavy metals.

The Eastern aquifer lies under the West Bank, which draws almost all its water from this source. Of the three, it is in best shape, though it too is being over-pumped.

The Coastal aquifer is in deep crisis. As a UN report in 2012 acknowledged, the Coastal aquifer has been overexploited for many years, since Egypt controlled Gaza in the 1950s. Nearly 95 percent of its water is now unfit for human consumption because of pollution from seawater intrusion, fertilizers, and sewage. Millions of unregistered and unauthorized wells have been sunk nevertheless — water has to come from somewhere — which exacerbates the problem. The report's conclusion was stark: Gaza may not be a "livable place" by 2020, unless it was remedied by "an enabling political environment," tactfully left unspecified; and as of 2014, Hamas and Israel showed no signs of such enabling.[9] Instead, Israel dismissed the report because one of its main conclusions was that Israel should abandon any blockading of Gaza in order to stimulate its economy and development. Israeli water authorities have offered to sell water to Gaza but only if the Palestinian Authority agreed — Israel does not deal with Hamas. Hamas has not responded, nor has Fatah.

There is plenty of desalinated water in Israel. Why not in Gaza? For more than twenty years, a major desal plant has been planned and even promised but never built. The reasons were partly technical — desal needs lots of power which Gaza doesn't have — but overwhelmingly political. No one can agree on where it should go, who should build it, or who should benefit. So nothing is done and Gazans must perforce buy expensive bottled water.

In Jordan, where 80 percent of all land is desert and only 5 percent considered arable, and where water riots have become an annual rite of spring, hopes are pinned on two large-scale water transfer schemes. In 2012, villages were getting water once a week from the national water system, and in some provinces it was worse than that, villages often going without for two weeks out of three. It surely doesn't help that water losses from rotting pipes averaged from 24 percent in some provinces to as high as 60 percent in others.

The first of the water transfers is the Disi Water Conveyance Project, constructed with World Bank and German Development Bank participation. (The second is the Med-Dead conveyance, already discussed in chapter 8.) It is designed to pump a hundred million cubic metres from the Disi aquifer, deep underground on the Saudi border, and take it to the capital, Amman, and thence to the northern provinces. The Amman branch went into operation in 2013; the northern route is supposed to be completed by the end of 2015. The cost is well over a billion dollars.

An independent study by Jordanian, Israeli, and American scientists found high radioactivity from many of the wells feeding the pipeline. The result was a flurry of reassuring press releases from the Jordanian authorities that pointed out that Saudis have been drinking from the same aquifer for years without apparent ill effect and that the Red Sea town of Aqaba has been using the same water for two decades without any "significant" increase in cancer rates. The water ministry, for its part, said not to worry because (a) the water could be diluted to bring it up to standard, (b) at any rate, the contaminant, radon, dissipated when exposed to air, and (c) the independent study was inaccurate. A former head of the Jordan Valley Authority, Dureid Mahasneh, said the radiation problem could be treated through a process called ion exchange, or by dilution, if water to dilute it could be found. Said Mahasneh: "We don't like the results, but can we handle it? Yes."[10] Nonetheless, the Jordanians had to drink the water. It was all they had.

Turkey and Its Neighbours

Its neighbourhood is a fractious one for Turkey. It shares borders with Iraq, a disintegrating state with ISIS troubling its heart; Syria, a disintegrated state with ISIS and a grab bag of rebel guerillas controlling parts of the country and the dictator, Assad, either on "our" side or not, depending on your perspective; and Iran, still a more or less theocratic autocracy, reeling from economic sanctions but a mortal enemy of ISIS and an unlikely ally of whoever will have it. And Turkey itself, of course, is still quivering from the aftershocks of the Gezi Park protests against Prime Minister Erdoğan's apparently boundless ego.

All of this is complicated by a troubling water crisis: the waters of Babylon are running dry.[11]

It is especially troubling here, for the Fertile Crescent, as it was once known, was where irrigation was invented by the Sumerians, creators of humankind's first urban civilization. They used the Tigris and the Euphrates Rivers as a source of nourishment and energy, and the crescent from Baghdad to the Gulf bloomed while they became rich in art and culture and political power. Eventually, they disappeared or were absorbed and were followed by the Assyrians and the Babylonians, and by the conquering Greeks and then the Romans, who were supplanted by the Ottomans and Islam. And throughout it all the bountiful high hills of Anatolia continued their nourishing flow. No one thought it would end.

But no one counted on there being almost 80 million people in Turkey or that drought would dry up the groundwater and diminish the rivers' flow. Or that climate change would make everything worse. Or that the Turks would attempt to re-engineer the two greatest rivers of their country, thereby forever altering their river-basin ecosystems and often depriving their downstream neighbours of the water they had taken as their right.

The Fertile Crescent is getting drier. For seven years starting in 2003, a pair of US scientific satellites measured the fresh water they found beneath them. The satellites, called GRACE, an oddly elegant acronym for Gravity Recovery and Climate Experiment, didn't measure the water directly but the effect water had on the patch of earth where it was found — or not found. "It's like having a giant scale in the sky," said one of the lead investigators, Jay Famiglietti, a hydrologist at University of California, Irvine. When GRACE peers downward, it can measure how strong the gravity is. Within a given river basin, rising and falling water reserves subtly alter the earth's mass, and thus how strong the local gravitational attraction is. By doing these measurements over time, changes in the water mass can be quantified and compared. "It is the only way we can estimate groundwater from space right now," Famiglietti said. With it, however, fairly close estimates can be made.[12]

GRACE found that from 2003 to 2010, the Tigris and Euphrates basins lost 144 cubic kilometres of water, an amount equivalent in volume to the Dead Sea. It is, Famiglietti told NASA, "An alarming rate of decrease in total water storage in the Tigris and Euphrates river basins, which currently have the second fastest rate of groundwater storage loss on earth, after India." Groundwater tables were dropping at thirty centimetres a year throughout the period measured. "The rate [of the drop] was especially striking after the 2007 drought. Meanwhile, demand for freshwater continues to rise and the region does not coordinate its water management because of different interpretations of international laws."[13]

The amount of water, enough to provide a steady supply for ten million people for a decade or more, was lost through over-pumping: in 2007 alone, the Iraqi government drilled a thousand new high-volume wells just to counter the shortfall in rain. No one knows how many more private landowners drilled wells of their own, but it was almost certainly just as many again. The abstraction rate has been more than four times the replenishment rate. And the water situation in the region will only get worse as climate change advances.

The Tigris and Euphrates systems are as stressed as the aquifers. All the governments concerned have built massive dams, containment basins, and networks of pipes. No comprehensive treaties or management documents have been signed, on the airy assumption that all would continue as before. This looks increasingly unlikely. As Famiglietti told the *Economist*, "The region is ready for collaborating on the science of water management. Whether it is ready for an international legal framework, I have no idea."[14]

Both rivers rise in the moisture-rich mountains of eastern Anatolian Turkey. The Tigris flows southeast, crossing low mountain valleys and the rolling Turkish plains, briefly becomes the border with Syria, and then heads, still southeast, through Iraq, passing through Baghdad and more or less skirting the Central Marshes before joining with the Euphrates. For its part, the Euphrates takes a rather more circular route, starting near the Black Sea and making a southwesterly curve through a series of lakes before crossing into Syria. From there it too heads southeast, crosses into Iraq, traverses the desert south of Baghdad, and skirts the Hammar Marshes to the south — or what would be the

Hammar Marshes if Saddam Hussein hadn't (mostly) drained them in his genocidal hunt for Kurds. For Syria, the Euphrates is the main water source. Iraq is dependent on both. Turkey, where the rivers rise, considers their water to be its property and stoutly maintains that it is for Turkey alone to manage their flow before it leaves the country, historical rights and prior use be damned.

This attitude has led to GAP, the Southeastern Anatolia Project, one of the most ambitious water basin development schemes on the planet and certainly the largest on an international river that is progressing without the say-so of the other parties. When completed, it will comprise twenty-two dams and an attendant network of irrigation canals, weirs, and barrages on a multitude of rivers that will irrigate 1.7 million hectares of currently non-irrigated land. Experts who have studied the project say that, when completed and while it is all filling, Syria's take from the Euphrates would drop almost 40 percent, and Syria currently does not have the resources or the energy to fight the Turks in addition to their many internal enemies. Iraq would lose almost all of its current Euphrates withdrawals, though at least it still has the Tigris to fall back on.

Syria's "unrest" (an awful euphemism to describe a country that is rapidly falling into a state of barbarism) is "in the most direct sense, a reaction to a brutal and out-of-touch regime," as a report for Washington's Center for American Progress put it. "However, that's not the whole story. The past few years have seen a number of significant social, economic, environmental and climatic changes in Syria that have eroded the social contract between citizen and government." By this the authors meant the drought that lasted from 2006 to 2011 and drove literally millions off their land.[15]

Those droughts are getting worse, and the country is ill-equipped to deal with them.

Egypt and the Nile Basin

Saudi Arabia, a country without a river basin of its own, may seem to have nothing much to do with the Nile and its basin, but it does. In an ongoing act of ecological vandalism, around thirty years ago, the Saudis decided they needed to be self-sufficient in staple grains such as wheat. To do this, they needed water, and the only water available was deep underground. Being expert drillers, they tapped into the aquifer and within a decade had a flourishing irrigated agricultural zone — and became self-sufficient in wheat, just as promised. But only for a while: by 2010, the water was gone and was not being replenished. So the Saudis turned to virtual water instead. Since they had money to spare, they started buying farmland in Ethiopia and Sudan, where they could grow wheat with someone else's water. Soon they found themselves in competition with the Chinese, who were doing the same thing.

In both Sudan and Ethiopia, there seemed to be water to spare — the Nile flows through both countries (and the Blue Nile rises in Ethiopia). But tapping into that water had another consequence: it diminished, or threatened to diminish, the Nile water available to Egypt. And the Egyptians really have no other source except for some fossil aquifers of their own. Egypt regards any threat to its Nile water allocation as a justified cause of war, and early in 2013, belligerent noises were being heard on all sides.

Those belligerent rumblings have a long history, dating back in some senses to the Pharaohs. But it wasn't until the British became the colonial masters of the Nile basin that they had any real focus — not until dams were contemplated along the Nile's length that would alter the flow and amount of water reaching downstream countries. And Egypt, of course, is last in line.

The British were also the first to contemplate treating the basin as a single entity. The precipitating event was a shortage of cotton on the world market and Britain's desire for Egypt and Sudan, then colonies, to pitch in to help remedy this. It would take perennial irrigation to bring off, and the Nile's traditional flood-fed methods just wouldn't do. Squabbles ensued inside the Colonial Office in London between Egyptian and Sudanese specialists as to where, exactly, this development should take place, upstream or down. By 1904, British engineers had reached Sudan's Sudd Marshes, where the Nile water pauses for a while before resuming its northward flow. The plan was to punch a canal through the marshes, thereby eliminating a lazy S bend in the river, and back in Cairo, plans were drawn up to regulate the river's full flow. To bring this off, they sought, vainly, to sign an agreement with the still-independent Ethiopians stating that the Blue Nile would not be tampered with. It wasn't until 1929 that they were able to impose the Nile Waters Agreement, which sought to apportion the Nile water among riparian countries and colonies. Egypt was guaranteed forty-eight billion cubic metres a year, out of an estimated average flow of eighty-four billion cubic metres a year. Seasonal variance was around 25 percent, and Egypt was also guaranteed all of this "timely flow," which meant that the Egyptian lobby in London had won. Sudan would be able to grow cotton only in the summer months, while Egypt could do so whenever it wanted. The treaty also guaranteed that no "works" would be developed anywhere along the river's flow without Egyptian acquiescence. The Ethiopians were not consulted and were not a party to this agreement.

After Egyptian and Sudanese independence, this treaty was revisited and a new agreement was signed in 1959, the Agreement for the Full Utilization of the Nile Waters, which essentially reaffirmed the earlier one. The river's average flow was still at eighty-four billion cubic metres a year, though it was now acknowledged that evaporation reduced this by ten billion or so, leaving seventy-four billion to be allocated. Egypt was still guaranteed its forty-eight billion, and Sudan only four billion, with the surplus divided one-third to Egypt and two-thirds to Sudan. Once again, Ethiopia was not consulted.

This treaty is still in force. Ethiopia has several times said it would go its own way and that what it did with the Blue Nile waters within its territory was no one else's business. Plagued with civil wars and then governed by a radical Marxist government that had no time for development, only repression, Ethiopia did little for decades. But the eventual settlement of Ethiopia's internal war and the destruction of its Derg tyrants raised alarm in Egypt. So did the fact that the

new government of Meles Zenawi seemed both confident and de-
termined — and willing to look to China for financing. Egypt under
Mubarak was able to bully Western financial agencies from financing
anything that would threaten its interests, but the Chinese proved
impervious.

Early in 2013, the Ethiopians began diverting a stretch of the Blue
Nile to allow construction of a massive six-thousand-megawatt hydro
dam (the Great Ethiopian Renaissance dam, to give the thing its grandiose formal title) near the Sudanese border, to vociferous protests
from downstream. The Renaissance dam is only one of many planned.
Sudan and Ethiopia between them have no fewer than twenty-five
new dams on the drawing boards, all aimed at protecting water security, generating electricity, and boosting food production. Egypt and
Sudan both complained that the dam would violate the Nile treaty that
essentially guaranteed them 90 percent of the Nile's water, ignoring
the fact that no Ethiopian signature was to be found anywhere on the
document. "The dam will cause no harm to anyone," Ethiopia's water
ministry declared. Egypt's president, the now-jailed Mohammed Morsi,
begged to differ: "As president of the republic, I confirm to you that
all options are open," he said. "If Egypt is the Nile's gift, then the Nile
is a gift to Egypt.... If it diminishes by one drop, then our blood is the
alternative." In return, Dina Mufti, spokesman for Ethiopia's foreign
ministry, said his country was "not intimidated by Egypt's psychological
warfare, and won't halt the dam's construction, even for seconds."[16]

In 2014, the belligerent noises mercifully subsided. There have
even been indications from the new military government in Egypt
that it is amenable to a significant shift in attitude, from considering
Nile water "theirs" to more or less accepting the idea that the Nile is
an African river with regional ownership and that cooperation would
benefit everyone. Egypt's president, Abdel Fattah el-Sisi, sent his foreign
minister to Addis Ababa, who returned to announce "a new phase of
our relationship, based on mutual understanding, mutual respect and
a recognition that the Nile binds us." A report on the dam's impact has
not been publicly released, but the Ethiopians have declared it supports

their notion of the dam's minimal impact. Water ministers from both countries repudiated Morsi's "all options are open" stance. "Previous statements were made in the heat of the moment," as Egypt's foreign minister put it. For his part, the Ethiopian foreign minister put it this way: "We have two options, either to swim or sink together. I think Ethiopia chooses, and so does Egypt, to swim together."

No water war, then. At least not yet. The good news is that the Nile basin, for all its potential for conflict, has been — and still is — a locus of cooperation and agreement. Sometimes reluctant, and sometimes cooperation with *sotto voce* rumblings, yes, but it still counts as a success.

Yemen

For more than a decade, the notion that Yemen may be the first country in the world to run out of water has been a cliché among hydrologists. A cliché but not a joke: the most optimistic projections are that the capital Sana'a will run out entirely in a decade, by which time its population will be peaking at more than four million. Where are they to go, these new water refugees? The forlorn hopes of the country's rulers rest on desalination, which is possible but for which there is no money, or — a more desperate scheme — on moving the capital to the coast. That would cost somewhere around $40 billion at best guess, and money is increasingly scarce — two-thirds of national revenue comes from rapidly diminishing oil reserves. Not that it would help much to move to the coast; water is short there too.

It obviously complicates matters that Yemen's governments come and go as reliably as winters, and that the country was overrun by a Shiite Houthi rebellion (the Houthi movement is named after its leader, Abdul-Malik al-Houthi) in early 2015, a state of affairs that devastated the country's major cities like Aden, expanded to a full-scale civil war, and threatened to erupt into a wider regional war between proxies for Saudi Sunnis and Iranian Shiites.

The country's central highlands, once the locus of a flourishing coffee industry, has seen its wells drying up and its remaining aquifers dropping sharply, leaving residents to fight for jerry cans of water from private water sellers, most of whom buy their water from illegal wells and resell it at grotesque markups. In the west of the country, along the coastal mountains, where there was once a flourishing agricultural zone that exported pomegranates and grapes, the Radaa basin aquifer has run dry and the country's most obvious export is now disaffected young men, recruits for al Qaeda and its more sinister offspring. Of the country's fifteen known aquifers, only two are anywhere near self-sustaining. What dams exist tend to leak, and when the rains do come, such as the heavy downpours and flooding in 2010, virtually none of the water is captured and stored. In some highland cities, the taps now run for only one day a month, and residents must store as much as they can in the time available — which has led to widespread water theft and revenge-motivated vandalism. The country's interior ministry confirmed in 2013 that water-related disputes were already causing more than four thousand deaths every year as predatory bands descended on ill-defended villages to steal what water they could find — far more deaths than occurred in the many al Qaeda attacks over the last few decades.

And yet 90 percent of Yemen's water is used for agriculture and almost half of that is used to grow khat, a mild narcotic and a known water hog. Khat has displaced what used to be food crops, causing food prices to spike, with resulting food riots.[17]

Wells are supposed to be a monopoly of the state, but sharia law, the basis of Yemeni jurisprudence, allows private wells on private property, so quasi-legal wells abound — about two-thirds of Sana'a's residents get their water from illegal wells, most of which have become contaminated as the city's sewage seeps into the groundwater. Government ministers, notoriously, all have deep wells in their houses.

There is nothing natural about Yemen's water crisis. It is almost entirely man-made.

India and Pakistan

On the face of it, water relations between India and Pakistan represent a triumph of diplomacy: the two neighbours, both nuclear armed and often much too ready for battle, signed a treaty in 1960 governing the water of the Indus basin, and despite three shooting wars, dozens of terrorism incidents, and seemingly unending violent rhetoric, the treaty is still holding.

It has been a near thing.

Water of the Indus basin is the main issue. The Indus is a river of considerable size — three times the size of the Tigris-Euphrates system and as big as two Niles. The basin is also the home of one of the oldest irrigation areas on the planet.

The partition of the country into Pakistan and India by the departing British also divided the Indus basin, Pakistan getting most of the irrigated lands and distribution canals, but with India, as the upstream power, controlling the flow. As well as the Indus itself, all three of the major tributaries, the Sutlej, the Beas, and the Ravi, flow from India into Pakistan, two of them forming the national frontier for several hundred kilometres. Since Pakistan depends on irrigation for 80 percent of its food, this arrangement is an obvious source of tension.

Independence from the British came in 1947, and the first belligerent action over Indus waters occurred less than a year later. The trigger was a unilateral action by the India state of East Punjab, which, finding itself short of irrigation water, cut off the flow to West Punjab in Pakistan, something it did without warning and without consulting its own federal government. The Pakistani prime minister, Liaquat Ali Khan, expressed his country's fury in a stiff telegram to his Indian counterpart, Jawaharlal Nehru, who replied in kind, his newly independent hackles up. The brouhaha continued until prudence was finally exercised and an international conference announced. A decade went by without a formal agreement, but at least the water was flowing again. Finally, in 1960, the International Bank for Reconstruction and Development, as the World Bank was then called, brokered an agreement, assigning the eastern rivers of the basin — the Sutlej, Beas, and Ravi — to India, and the western rivers, the Indus itself, the Jhelum, and the Chenab, to Pakistan.

This is the treaty that is still in force, sixty-four years later.

Despite this long history of cautious non-belligerence, cross-border amity started to fray again in the second decade of the new century. There were many causes but chief among them was ongoing drought — and India's determination to build run-of-the-river hydro schemes on two of "Pakistan's" rivers, the Chenab and the Jhelum. Allegations soon followed in the Pakistani press that India was bent on stealing Pakistan's water, and, at the same time, accusing India of causing severe flooding along the Chenab itself. In a long interview with the newspaper *Dawn*, the Pakistani commissioner for Indus waters, Mirza Asif Baig, sought to tamp things down. No, India is not stealing water — "Some pseudo-water experts are spreading this notion. India will never do so because it would lose face globally. Our own response will not be subdued if it happens." Yes, less water was flowing in the Chenab, he said, but only partly because India has constructed new irrigation zones — Pakistan has done the same thing. Yes, India is building hydro schemes on the rivers, but this is allowed under the treaty — it was the design of the project, not the projects themselves, that was causing difficulties. Yes, India built a dam called Kishanganga to which Pakistan objected — but international arbitration came down on the Indian side. No, we shouldn't scrap the treaty and seek a newer one. That would create too much uncertainty. A few tinkerings here and there wouldn't hurt, though.[18] His federal counterpart, the state minister for water and power, was not so positive. Chaudhry Abid Sher Ali directly blamed India for Pakistan's shortages — too many dams and hydro projects on rivers flowing into Pakistan, he said, could spell catastrophe for Pakistan. With water shortages looming, he urged his government to start building more dams of its own.

The real reason for mistrust, though, had nothing really to do with India and everything to do with Pakistani mismanagement, itself driven by corruption, sectarian violence, tribal squabbles, and increasing poverty. In the last forty years, Pakistan has built only two large storage dams, and those are already in trouble through excessive silt. In the same period, India built four thousand new dams — and China an astonishing twenty-two thousand. Pakistan's few dams store only about a 30-day supply; India's, a 120-day supply. No groundwater recharge schemes exist in the country, though it is well known that Indus water will likely diminish over time because of less glacial melt. Almost three-quarters of water used for agriculture flows in only three months of the year; without dams, the rest runs unused to the sea. Farming still employs about half of all Pakistanis and accounts for about a quarter of GDP, yet no plans exist for water conservation. It is still common for those Pakistanis who do have access to good piped water to leave the taps running all night because they find it soothing. Overall, Pakistan has only one thousand cubic metres per capita per year to spare, about the same as Ethiopia, down from five thousand cubic metres in 1947, and is defined by the Asian Development Bank as one of the most water-stressed countries in the world. A rapidly growing population is partly to blame for the reduction — there are already 190 million people in the country, two-thirds of them under age thirty, and the number will likely rise to 256 million by 2030.

A Pakistani water expert, Munawar Sabir of the University of the Punjab, had this to say: "Our agricultural input has decreased, annual floods have become routine, and in 2013 alone, more than 178 people have been killed. The infant mortality rate is high because of contaminated water. Water resources of both [India and Pakistan] are eventually sharply depleting, amounting to dangerous levels."[19] Water pollution makes it worse: the capital, Karachi, now contains about eighteen million people, and more than half of them are obliged to drink unsafe water. Typhoid, cholera, dysentery and hepatitis are common.

The Pakistani newspaper *Dawn* had this to say in an editorial in 2013:

> This [water crisis] has led many — from farmers to opposition politicians to ministers to jihadi groups — to blame India ... for Pakistan's water crunch. ... It isn't without reason that some experts have warned of water wars in South Asia, one of the world's most water-stressed regions. ... The country's population is predicted to double by 2050, meaning that the people will have access to just half the water in 2050

they have now even if they start using the available resource efficiently and climatic changes don't reduce flows in the Indus river system.... The situation can still be salvaged. But it'll require efficient use of water, the development of more storage capacity, resolution of provincial water disputes as well as engagement with India to find a peaceful solution to trans-boundary water-sharing.... Unless effective actions are taken now, the future appears grim.[20]

Punjab University's Sabir thinks it may not be quite that grim. Enough water exists, if it is used carefully and properly conserved, he says. Michael Kugelman, a fellow for South and Southeast Asia at the Woodrow Wilson Center, adds: "The root of the problem ... [is that] water is often misallocated and wasted. This all gives the illusion of scarcity. In both countries, better demand-side management would resolve the water crisis."

But better management may be hard to come by. Especially when Jamaat-ud-Dawa, the Pakistani militant group that was behind terrorist attacks in Mumbai in 2008, issues a statement against "India's water terrorism," and declares that "water flows, or blood."[21]

India and Bangladesh

No fewer than fifty-four rivers are shared between India and Bangladesh, and all of them are stressed. The main ones are the Ganges-Brahmaputra system and its tributaries, especially the Teesta, the subject of increasing acrimony.

Here's the basic geography: most of the flow of the Ganges, holy river of the Hindus, originates in India's Himalayas, but about a third comes from Nepal and China. It traverses the Ganges flood plain, briefly becomes the border between India and Bangladesh, and passes wholly into Bangladesh, where it is renamed Padma. The Brahmaputra rises in China and circles around eastern Bhutan before briefly passing through India and into Bangladesh, where it is called the Jamuna, before joining the Padma and emptying into the Bay of Bengal. The

Teesta, for its part, rises in the Indian territory of Sikkim and flows through the northern fringes of West Bengal before crossing into Bangladesh and merging with the Jamuna. Jointly, the basins of the three systems comprise only a fraction of one percent of the earth's landmass, but they contain about 10 percent of the planet's population. The plains are wonderfully fertile, the rivers consistent, yet most of the people are poor.

The politics are messier than this straightforward description would indicate. Just as with Pakistan, disputes over the river-basin allocations date back almost to independence, in 1947, when Bangladesh was still the Eastern Province of Pakistan. The Indians, again, started it. Their port city of Calcutta, as it was then called (now Kolkata), was in danger of silting up because not enough water was flowing down the Hooghly River to flush the silt into the sea. India therefore said it would build a barrage across the Ganges in the Bengali region of Farakka, close to the border with Pakistan. From there, water would flow through a canal into the Hooghly, desilting Calcutta's harbour, or so it was hoped. East Pakistan, dependent on the two main rivers for water, inevitably protested to no avail. The Indians went ahead anyway, and construction started in 1962.

Three years later, the two countries fought the first of three Indo-Pakistan wars, complicated by the rise of a vociferous separatist movement in East Pakistan. In 1971, Pakistan's Eastern Province became the new country Bangladesh, and since India had been covertly helping the separatists, it seemed that an amicable settlement of the Farakka dispute would follow.

It was not to be so simple.

The rise of Islamic militancy on the one hand and Hindu fundamentalism on the other meant that for domestic reasons, neither side could afford to be too accommodating. Both sides stalled and it wasn't until 1974 that a cautious "interim" agreement was signed, allowing India to build its barrage to divert around four hundred cubic metres per second into the Hooghly and thence to Calcutta. Both sides, it turned out, hated the agreement: India because it didn't get the a thousand cubic metres Calcutta said it needed, and Bangladesh because its leader, Mujibur Rahman, appeared weak and overly concerned with Indian sensibilities. Rahman was assassinated a few years later and the water agreement was thought to be the main cause.

For another decade, not much happened. Finally, in 1992, the two governments sent delegations to a conciliation meeting, but both sides departed angrier than they had arrived. India's Farakka diversions continued. In 1996, a Bangladeshi government report measured the flow at Hardinge Bridge and found it had shrunk from 1,740 cubic metres per second to only 362, drying up the Gorai River on which southwestern Bangladesh depended for its water. It was a surprise to everyone, then, when both countries signed a formal Ganges Water Sharing Agreement in 1997 that laid out a complicated water-sharing regime whose key was to ensure that Bangladesh got a minimum of 833 cubic metres per second in the driest part of the year, from March to May. It was supposed to be a thirty-year agreement, meaning it would have to be renegotiated in 2027.

The Teesta River, though, has complicated things. A separate agreement had been reached over its water in 1983 in which India was allocated 39 percent and Bangladesh 36 percent of the river's flow. A formal treaty giving each side 48 percent (the rest was for the natural flow) was supposed to be signed in 2013, but there was a hitch: Mamata Banerjee, the newly elected chief minister of West Bengal State, balked — the river's flow had been sharply reduced through a recent drought, and in her view sending water downstream to the Bangladeshis would unfairly penalize her state. Besides, Sikkim, where the river originated, had already built five dams on its upper reaches and was planning thirty-one more, making things incrementally more precarious. Banerjee was sharply critical of her own federal government. At a rally before her election, she declared that "the central government is giving away the Teesta water, depriving the people of West Bengal of their rights," and urged everyone to "teach them a lesson at the ballot box."[22] Of course, they *did* teach them a lesson, throwing out the Congress Party and electing Narendra Modi, himself a former state first minister (Gujarat), a Hindu nationalist and a strong believer in states' rights, which in India include water. By 2014, it still seemed unlikely that Modi would want to jeopardize relations with a strongly nationalist region by reaching an agreement with foreigners — and Islamic foreigners to boot. When he was in New York in September 2014, Modi met with his Bangladesh counterpart, Sheikh Hasina, and was reported to have said he would "look into it" and that it would be "just a matter of time." "Looking into it" will almost certainly have to wait until 2016, when West Bengal goes to the polls again.[23]

Despite all this, in non-water matters, relations between the two countries are as cordial as can be expected and hostilities are not likely.

One good sign is that several companies have set up tourist cruises on the section of the Teesta River that passes through Assam, once the locus of violent insurgencies. "This is an India very few [outside] people have seen," Nirmalya Choudhury, manager of operations at Assam Bengal Navigation Company, said, with commendable understatement. "The pleasure of meeting local people who welcome foreigners with an open smile cannot be explained in words."[24]

China and Its Neighbours

As discussed earlier, China is planning a multitude of new dams on the Salween River, called the Nu in China, without consulting its riparian neighbours, mostly Myanmar and Thailand. The Salween, at 2,413 kilometres, has so far managed to remain the longest un-dammed river in Southeast Asia. That's about to change, and not just on the Chinese side: both Myanmar and Thailand, with Japanese development help, have undertaken "feasibility studies" for new hydro projects, and development pressure in the basin, home to and livelihood for thirteen million people and thirteen ethnic groups, is edging upward. No formal agreement on the Salween exists, and experts say the development plans of the three nations are not always compatible. China, by far the largest, richest, and strongest country, is acting unilaterally, threatening to divert as much as 10 percent of the river's flow for its own purposes, and there are fears that release of sediment-rich water from the dams would damage fisheries and erode banks where farmers cultivate rice.

Downstream, there is somewhat more amity. The Thais need water — the northeast of the country has been suffering shortages, caused mostly by deforestation, and the country has essentially dammed all the rivers it can. The Myanmar military junta has indicated it wants to cooperate with Thailand, mostly because it needs more electrical power and wants Thai money to help it acquire what it needs. The mechanism is a set of five hydro dams just short of the Thai border; studies say they could produce thirty-four hundred megawatts of electricity, a quarter of which would be allocated to Myanmar, with Thailand purchasing the rest at an agreed price. There has been ethnic unrest in the area — the Myanmar military has been hounding the civilian population in Shan State, home to a militant insurgency violently opposed to the government in Rangoon, and the dams would be vulnerable to sabotage. But the agreement, with the Thais clearly holding their noses, stands pending financing.

Actual water conflict is unlikely. Growing water stress, peasant revolts, and ethnic strife are more probable.

The Aral Basin

When in the fall of 2012 the unloved leader of the gas-rich but water-poor republic of Uzbekistan threatened his neighbours, also rump states of the former Soviet Union (Afghanistan, Kazakhstan, Kyrgyz Republic, Tajikistan, and Turkmenistan) with armed conflict, he was carrying on a lamentable tradition dating from the dying days of the USSR. Still, Islam Karimov was unusually forthright: Uzbekistan, central Asia's most populous country, depends for its precious water on rivers that rise in Kyrgyzstan and Tajikistan, and anything that threatens that resource would precipitate conflict, he declared. He was referring mostly to his upstream neighbours' desire to revive the massive hydroelectric schemes proposed by Soviet planners and largely abandoned in the aftermath of the USSR's dissolution. In particular, he singled out Tajikistan's Rogun dam, a multibillion project on the

Vakhsh River, itself a tributary of the Amu Darya. It was "started" in 1967 but remained unfinished in 2014, pending a slowly moving World Bank assessment. If it is ever completed, it will be the world's highest dam, a full 335 metres of rock and rubble and concrete slurry built in an earthquake-prone region with millions living downstream. (Not much smaller is the already completed Nurek dam, on the same river.) As reported by Reuters correspondent Raushan Nurshayeva, Karimov was sarcastic and caustic: "They're going for the Guinness world record, it would seem, but we're talking here about the lives of millions of people who cannot live without water. These projects were devised in the 70s and 80s, when we were all living in the Soviet Union and suffering from megalomania, but times change." Then, in a nudge to World Bank assessors, he suggested that "hydropower structures today should be built on a different basis entirely." Tajik government officials, said Nurshayeva, "were not available for comment."[25]

This sorry tale starts with Soviet planners in the Cold War era, when desk-thumping Nikita Khrushchev promised he would "bury" the West by outperforming it in every way, including in the very area for which the Soviet Union was most laggard, food production. To do this, the USSR was going to create a new world breadbasket in the Virgin Lands of central Asia.

Instead, it caused the ecological catastrophe that is the Aral Sea.

Or rather, *was* the Aral Sea, because the Virgin Lands were ill-suited to any kind of large irrigation schemes, as would soon become painfully clear.

Not that anyone cared at that point. The Aral is fed by two rivers, the Amu and the Syr, which rise in the Tien Shan and Pamir mountain ranges of the Himalayas and flow northward through their alluvial valleys and the Kara Kum and Kyzyl Kum Deserts before emptying into the Aral, one into the southern tip and the other into the northern. These two rivers are the only real source of fresh water in a system that comprises one and a half million square kilometres and (in 2010) somewhere around sixty million people. In 1956, the Soviet agricultural commissar opened the Kara Kum Canal and began diverting water from the Aral. Soon, both rivers were diverted into the new farming zones in the Uzbek and Kazakh deserts, irrigating more millions of hectares, and the flow into the Aral dropped by more than half and then stopped altogether.

What happened subsequently is well enough known. The little fishing village of Muynak, soon to be a poster child for ecological horrors, was an island in the Amu Darya Delta in 1956. By 1962, it was a peninsula. By 1970, the sea was ten kilometres away and the village was abandoned. Ten years later, the nearest water was forty kilometres away, and by the turn of the millennium, it was seventy-five. In 2014, to complete the mess, the eastern basin dried up completely. Half-hearted attempts by the Uzbeks and the Kazakhs to resurrect yet another Soviet conceptual folly (turning the great Siberian rivers southward to water the Aral basin) never got any traction.

The Aral's most recent chapter began in 2005, when the World Bank, with some reluctance but under strong pressure, granted Kazakhstan a $68 million credit to build a thirteen-kilometre dam across the northern tip of the former sea, thereby cutting it into two lobes, a small Kazakh one to the north and a larger Uzbek one to the south. At the same time, the Kazakhs restored some of the Syr River flow and their small part of the Aral began slowly to fill. By 2008, miraculously, a small fishing industry resumed as the salinity levels in the northern lobe dwindled. To the south, however, the Uzbeks had no water to divert — it was all still being used to irrigate the cotton industry, one of the country's major exports.

Tensions remain. The Uzbeks, with no water of their own, have a finely tuned sense of water paranoia (they have on occasion accused Tajikistan of exporting disease in the guise of fresh water, a notion encouraged by Tajik bandits who have raised the possibility themselves). Early in this century, the Uzbeks admitted to hatching a scheme to invade Turkmenistan in an effort to control its water. Mostly, these tensions have subsided into low growling but not much else, except over the Rogun dam.

On a positive note, if agriculture is fixed (shifted to less water-hungry crops) and infrastructure improved, there is water enough for everyone — the region is well above the sixteen hundred cubic metres per person that is the benchmark for water stress — indeed, more water is available per person than in Denmark, for example. The countries in the region remain among the highest users of water anywhere: a Turkmen consumes four times as much water as an American, and thirteen times more than a Chinese — leaving obvious room for improvement. This wastefulness is not helped much by grandiose projects such as Turkmenistan's Golden Age Lake, which will draw already scarce water from the Amu Darya to fill a "lake" in the desert as "a symbol of national pride."[26] Still, there are some signs of improvement. By 2012, water partnerships and local committees had helped raise water productivity by 30 percent. If that progress is extended, the region could become a model similar to that of the Nile, with amity prevailing instead of conflict, promises instead of threats. Uzbekistan could sell gas for energy to Tajikistan instead of threatening to cut it off; the Tajiks could, in return, ensure downstream water supplies.

Good sense could well break out.

The Kura-Araks System

Another leftover squabble from Soviet times involves the Kura-Araks River basin in the South Caucasus region between Georgia, Armenia, and Azerbaijan. The main issue here is pollution, rather than supply. A report on the region suggested that progress in arriving at water-management programs has been slow, "but a foundation is being established for future work between the nations when they are ready." Don't hold your breath. Armenia and Azerbaijan have been in a state of armed hostility since 1988, and even more so since 1991, when bloody conflict erupted over the disputed Nagorno-Karabakh region, leading to a land grab by Armenia that still holds, despite a 1994 ceasefire. It is hard to talk about water when land is the dominant issue.

The Azeris, as the Azerbaijanis are called, lack groundwater resources of their own and are dependent on the Kura-Araks River flow, and complain frequently about pollution from upstream neighbours entering the country. No one pays any attention, but the pollution really is severe. The Kura-Araks is pregnant with chemical, industrial, biological, agricultural, and even radioactive pollutants, and what wastewater treatments once existed have collapsed.

Still, early in the century, a meeting was held in the Georgian capital, Tbilisi, at which representatives from all three governments showed up, along with NGOs, environmental agencies, scientists, and EU facilitators, who all agreed to work toward a common approach. The non-binding resolution signed by all concerned "took into consideration" notions such as supporting a culture of sustainable water use, improving exchanges of information about water, developing an environmental strategy about hazardous materials being dumped into the river, and developing a transboundary water plan.

There are copies of the resolution in all three capitals.

It's a start.

By the standards of these watersheds, water conflicts elsewhere are minimal, and the flashpoints a lot less flashy.

Measured against the problems in Yemen, the Middle East, central Asia, and much of North Africa, Europe's water stresses are moderate, at worst. This is not to say they are absent, and hydrologists are now commonly using the phrase the "Africanization of Europe" to denote increasing shortages and increasing risks of conflict. It's a term that annoys politicians in Europe's relatively arid south who thinks it exaggerates their problems, and also Africans who resent the use of the word "Africa" as a synonym for awful.

It's true that a third of European countries have relatively low availability (less than five thousand cubic metres per person per year), with the most stressed countries in the south; it is true that 20 percent of European countries depend on water from a neighbouring country; it is true that 140 million Europeans live in or near areas of groundwater overexploitation (and that two-thirds of Europeans depend on groundwater for their daily use); it is true that the average losses from creaky infrastructure are around 30 percent, and much higher — up to 70 percent — in many urban areas; it is true that 5 percent of Europe's wetlands have disappeared in just the last few decades; and, of course, it is also true that Europeans have waged war on each other as long as there has been a Europe. But still, Europe has the money and the politics and the technology to mitigate stresses where they appear, and in most places they are also getting policy right, pricing water for conservation. Various European agencies are beginning to contest industry's right to extract as much water as it wants to use — not through rationing but through pricing. Some farming will probably face the same pressure in the near future.

What conflicts have existed tend to be internal and tend to be in the south, where water is scarcest. In the south and southeast of Spain, for example, along the Costa Brava and the coast around Murcia, there have been clashes between farmers, whose water has been diminishing, and the developers of the golf-course-strewn, swimming-pool-laden resorts built for the holiday benefit of northerners. This has been complicated further by political payoffs and corruption (golf-course owners around Murcia have been declaring their fairways a "crop" in order to qualify for an agriculture allocation, Elizabeth Rosenthal reported in 2008).[27] As we discussed earlier, Barcelona, which really isn't in the south, had to briefly resort to tankering in water to help fill a reservoir, a futile exercise soon abandoned. Already parts of Spain are turning to desert. Groundwater has been depleted by hundreds of illegal wells, and water transfers from the north have been reduced because the north too is drying up. And it hasn't gone unnoticed that Spain has only a third as much water per capita as Portugal, which gets much of its water from rivers that cross the border from Spain.

Still, even in Spain there has been progress. Barcelona, spooked by the shortages in the recent drought, has improved its water efficiency. Madrid even more so — consumption in the capital has dropped 25 percent through better water conservation, saving (as the European Environment Agency points out approvingly) the equivalent of a reservoir providing over one hundred million cubic metres a year.

For the rest, most of Europe's water issues stem from pollution, not supply, and we've already discussed those. And other issues that are beyond the purview of this book: rising sea levels threatening low-lying areas, for example.

In North America, apart from the purely internal issue of confronting shortages caused by careless overuse and drought, transboundary issues are minimal. The two residual nagging issues are allocating the shrinking waters of the Rio Grande on the Mexican-US border, and, along the Canadian border, the vexed notion of America's perceived need for Canada's water, as mentioned briefly in Chapter 8.

In neither place will there be a water war.

Andrew Nikiforuk, a skillful and thoughtful journalist and observer of matters environmental, had this to say in a 2007 paper: "Make no mistake, Canada's water — through diversion, transfer, sale, trade or all of the above — is on the negotiating table in Canada/US relations. While water is not necessarily the top item of negotiation, and at times is dormant as an issue, it is there. In the long-term agenda within the context of freer trade and increased North American integration, Canada's water is up for grabs. As long as its status as a negotiable resource remains unclear, pressure to access Canada's water will continue to grow ever stronger."[28]

Because he *is* a thoughtful journalist and not prone to hyperbole, one must perforce take him seriously. Nevertheless, this notion that Canadian water is somehow up for grabs is overwrought, and wrong in its essence.

It is wrong for two reasons. One is the erroneous notion that "the Americans" speak with one voice. They don't. It's not hard to come up with examples of Americans who think Canada's clinging to water ownership is peculiarly regressive (the authors of *Water Markets*, for example, arguing for "freely tradable water rights" across the US-Canada border, which "will be accomplished best by a careful relinquishment of national or state sovereignty sufficient to create rights enforcement institutions which are free from the distorting influence of national, provincialism and political competition"). Well, good luck with that. The same writers seem puzzled that Canadians seem "willing for forego possible benefits of water trading opportunities that may be available under NAFTA." This goes to show the degree to which policy wonks can be remarkably ignorant of political realities.[29] But there are plenty of people, many of them influential and in positions of authority, who subscribe to the basic ecological principle that "water should be kept within its natural basin, treated with respect, and used efficiently" (a phrasing suggested by Adèle Hurley, director of Toronto's Munk School of Global Affair's Program on Water Issues, in her introduction to Nikiforuk's paper). The second reason is that the United States has enough water of its own despite the drought ravaging the Southwest, and that is not likely to change: the Pacific Northwest and the Atlantic Northeast are both likely to get wetter as climate change advances and so are parts of the Midwest. You might say that the American Southwest would find it easier to bully British Columbia into sharing its water than, say, Oregon, but either one is politically improbable.

At the other end of the United States, matters are worse. The Rio Grande has been in trouble for decades and is not improving. It is the fifth longest river in North America; from its source in the Rockies, it travels a little more than twenty-nine hundred kilometres and drops around three kilometres before emptying into the Gulf of Mexico. It passes through three states in the United States (Colorado, New Mexico, and Texas) and five states in Mexico (Chihuahua, Durango, Coahuila, Nuevo León, and Tamaulipas). From El Paso to the Gulf of Mexico, a distance of about two thousand kilometres, the river is the international frontier. Its watershed is massive, about 870,000 square kilometres, about 11 percent of the area of the continental United States. But for a good part of its run, it passes through the Chihauhauan Desert, where there are no tributaries to replenish it, and intense demand has over the years depleted it until, by the year 2000, the river no longer flowed all the way. American cities and users have extracted every drop from the river by the time it reaches El Paso, Texas, and the river doesn't resume flowing until the confluence with the Rio Conchos, four hundred kilometres downstream. The Conchos and its tributaries rise in Mexico, and so the lower reaches of the Rio Grande consist almost entirely of Mexican water.

NAFTA made things worse. Instead of locating industry in better-watered regions to the south, Mexico concentrated its efforts in northern areas close to the US border, leading directly to the industrialization of deserts, a folly as great as California's. As we have seen, most of the country's aquifers are already seriously depleted and the country as a whole, despite the relatively verdant south (and Yucatan), is under high water stress.

The stresses are going to get worse. But there won't be a water war. Both sides know that would be futile.

Part Five

Solutions

13
Reframing the Debate

About a decade ago in Windhoek, the capital city of Namibia, a water manager pressed into my hand a tumbler of recycled sewage and urged me to drink it. Which, of course, I did; it would have been churlish not to, for the water manager was endearingly upbeat about what he had wrought. I did the little ritual wine tasters do — swirling the water around in the glass as though it were the finest pinot noir, tipping it sideways to check for colour and turbidity, sniffing it gently for aroma, then pulling it through my teeth in that irritating wine-taster's dribble, to best get a feel for its flavour. Not to much avail: the water was perfectly clear, smelled of nothing at all, and tasted like water, nothing more. I drained the tumbler and felt much better — it was really hot outside and I'd been thirsty all morning.

I drank the water knowing what it was because I had just toured the treatment plant where it was produced and knew it had been built with German technology by Veolia of France, funded by the World Bank, and managed by well-trained local engineers, and I had no qualms whatever about its safety. I already knew that Windhoek was the first city, anywhere, to fully recycle all its water, whatever the source, into drinking water and that this was a good thing. Namibia is a water-stressed country with meagre rainfall and no rivers wholly its own. Water thrift had become essential as the population swelled. And water thrifty the country is. Namibia is not wealthy and is still in the early stages of development, but it is reasonably well governed and has no reservations about private-public partnerships as long as they are useful and affordable. It didn't have to be bullied by the World Bank or anyone else. It got there entirely on its own.

If it can be done here, I thought at the time, it can surely be done elsewhere. The technology is available, easy to replicate, and affordable.

A month or so later, back home on the Atlantic shore of Nova Scotia, I filled another tumbler of water, this time from my own tap, and did the same wine-tastery thing, trying to bring back from memory the sensations I had felt in Namibia. When I sipped the water, I could detect no difference from the earlier glass, either in clarity or taste. My water came directly from a well outside my house, unfiltered and untreated, and contained no sewage, treated or untreated (something we know, because we have the water tested each spring). If we really wanted to, we could do the same thing Windhoek was doing. The technology to turn the output of our own private septic system into a small on-site "membrane bioreactor" that would alchemize it into drinking water is available, if not yet cheap. But we have no interest in doing so, mostly because we have no need to. Our household has plenty of water. We live in an area of abundant rainfall, and we share a small peninsula with just a few other families. We conserve water anyway. It is a habit. We have four rain barrels to irrigate our small garden on the rare occasions it is needed, and use low-flush toilets and all the rest; our consumption is minimal, our extraction from the aquifer as low as it can comfortably be. We don't have to go this extra mile; we just do. Even though the water we save through our careful water use cannot help people in other places. Certainly not the thirsty citizens of Windhoek.

This discrepancy — between my situation and Windhoek's — was in my mind as I read a thoughtful piece in the journal *Nature* written by a researcher at MIT's Department of Earth, Atmospheric and Planetary Sciences, titled "Water Politics Must Adapt to a Warming World." About half way into his argument, Moshe Alamaro said this: "The reality is that there is simply not enough fresh water for everybody to use as much as they want, whenever they want."[1]

It sounds plausible, even anodyne, but it is not universally true, and framing it as he did is fundamentally misleading. And in an indirect way, actually damaging.

This notion of "not enough fresh water" is accurate, clearly, in California and Texas and in those parts of Brazil also suffering from drought, in the North China plains and in Israel and Syria and a multitude of other places. It is true everywhere aquifers are being overdrawn and where rivers are being overexploited, which means it is true in many places on this earth. But it is not true everywhere. Many places have more than enough water. Norway, for example. The Norwegians could easily reduce their consumption — but if they did so, it would benefit no one but themselves. Iceland has plenty of water, but it too could easily cut back — but if it did so, how would that benefit the parched towns of the American Great Plains? Everyone in Quebec could instantly give up drinking bottled water — but that would not compensate in any way for the reduced flow from glaciers in western Canada.

Alamaro's point is misleading in this way: There are two sets of water crises around the world, the crisis of supply (availability) and the crisis of contamination (pollution) — these, remember, are what the World Bank called the "dismal arithmetic of water." Virtually every country and most regions suffer from at least one of these crises, and sometimes from both. Infrastructure is crumbling almost everywhere, in rich countries and poor ones. Groundwater is shrinking alarmingly in dozens of places. But there is no single global water crisis that can be solved globally, with UN-style treaties and massive international action. Instead, there are many hundreds of local water crises, and all of them have to be solved locally.

It is counterproductive, therefore, to think of shortages "everywhere," or to maintain that the world's water problems are inevitably catastrophic for the planet as a whole. They are not. Yemen's water problems are catastrophic for Yemen, Pakistan's for Pakistan, Mali's for Mali. Mali's problems are not catastrophic for Haiti, or Bolivia or

Lubbock, Texas. The shrinking Rio Grande cannot be filled by water thrift in Denmark; polluted rivers in China cannot be scrubbed by America's Clean Waters Act. To think of all these as just one large problem, or even two large problems, makes the whole issue seem impossibly difficult to solve.

This is not to say that true global crises don't exist — in sectors other than water. God knows we have enough such global problems already without adding to the list. Climate change is, self-evidently, a global problem that can *only* be solved globally. Air pollution too: China's many hundreds of new and dirty coal-fired generating plants affect everyone on the planet — mercury from Chinese emissions has been detected in Oregon snow melt and so has nitrogen oxide and sulphur. And so when China and America agree to limit emissions — despite the scoffings of critics from both sides of the political spectrum — it benefits everyone on the planet.

Water is different. Nothing anyone else can do will increase Syria's supply of desperately needed water. Only Syria can do that — with our money if necessary, but the work will have to be local. Thrifty consumers in Philadelphia won't cause a single California tap to resume its flow. No amount of conservation anywhere will increase anyone else's supply. That has to be done where the supply is critical, nowhere else. Water problems are local problems.

That means they need to be fixed locally. Which means they *can* be solved locally. Which means "the water crises" are amenable to solution because they are smaller than we thought. We just need to approach them differently and design solutions appropriate to local conditions.

Charles Fishman, in his engaging romp through water politics called *The Big Thirst*, makes the same point persuasively:

> How you shower or water your lawn has no impact on the water availability of people an ocean away, and may not have any impact on people a time zone away. That doesn't relieve us of responsibility for water behavior and water habits — precisely the opposite. It remains we must take

responsibility for our own water issues, because no one else on the globe will. No one else can.

That's actually the good news about water: the problem is right with us, but so is the solution. Part of what is so frustrating about truly global problems like finance and climate is that even if you and your community, even your country, behave with thoughtfulness, discipline and fore-sight, if your neighbors don't, all the good work you do can be instantly undone. That isn't true of water. If Perth, Australia, remakes its water in an intelligent, sustainable fashion, that effort cannot be undone by water carelessness in Melbourne, Australia, or Istanbul.[2]

Despite this, there is at least one way that the global hydrological cycle does need to be considered globally though its impacts are still local. The hydrologist Robert Sandford puts it this way: "Changes in the extent and duration of Arctic sea ice and northern snowpack and continental snow cover have begun to have a cascading effect on weather and climate, right down to the mid-latitudes. The loss of relative hydrological stationarity is a societal game changer. It means that simply managing water in ways that are useful to us will no longer be enough. We now have to be alert to changes in the larger global hydrological cycle, and if possible, try to manage and adapt to them."[3]

And there are ways in which local water problems do have ripples beyond the water basins in which they are found. They still have to be fixed locally, but their consequences are often transnational.

One such consequence, as we have seen, is the way that water stress can cause water conflict. Water refugees are already straining international aid agencies, and their numbers will swell. Even those who are not refugees, those who stay home, are likely to be caught up in desperate efforts to get enough water to stay alive. This in itself has two negative effects: foregone human potential as human effort and ingenuity are diverted into mere survival, and growing anger, leading to growing militancy with resulting militarization and, increasingly, terrorism. On a larger scale, an example of militarized chaos is Syria; the revolt against the Assad autocracy wasn't caused by water shortages, but water stress was one cause among others — it exaggerated the social disruption and still draws it out as hundreds of thousands are driven off their land because their wells have run dry.

Solutions envisaged by the "water establishment," the World Water Council, the World Bank, and the International Monetary Fund make any rethinking of water problems rather more difficult than it might otherwise be, not because these institutions don't understand how local water issues are but because the solutions they propose are often grandiose infrastructure projects, much more costly than they need to be. In many cases, their definition of local isn't local enough.

I want to make one last point in this reformulation of water from global to local issues. As Fishman points out, abundance doesn't absolve water-happy consumers from a responsibility to conserve water, or to use it efficiently. I could use as much water as I wanted without disrupting the hydrological cycle in any way and without depriving anyone else of the water they need. But even my small well uses an electric pump to distribute the water through my house, and that electricity is generated in our province largely by fossil fuels. So it is not without cost. And multiplied by millions, it has a real negative ecological effect that is, indeed, global in scope.

Further, water problems may not seep across borders and into other jurisdictions, but attitudes do. And an attitude of entitlement, of profligacy, and of waste is surely contagious. *That* is a global problem.

14

The Way of the Engineer

It all seemed so easy, in those innocent days before we knew the globe was warming and before anyone knew where Three Mile Island was, never mind Chernobyl or Fukushima. We would build a string of nuclear power plants along the coast (the energy they produced would be so cheap, we wouldn't bother to meter it) and use it to strip the salt from seawater, and our water and energy issues would be solved. That was before we knew that a medium-sized nuclear generating plant would cost better than $3 billion to build and just as much to decommission, before the public developed its largely irrational fear of nuclear energy, and before anyone seriously started running out of usable fresh water anyway.

We know more now, about how energy and water issues intersect, about climate change and resultant droughts, about unforeseen consequences and collateral damage. We know more about how "easy" supply makes for heedless development. We probably don't know enough, but enough to make us wary of the promises of engineers. Which is a pity. Because the way of the engineer is one way out, even now.

The way of the engineer — what Peter Gleick calls "the hard path" — relies on dams and reservoirs, pipelines and treatment plants, and often centralized water departments and agencies to bring fresh water in (or create it on the spot) and take waste water away. "Once easy sources of raw water are captured, however, this path leads to more and more ambitious, intrusive, and capital-intensive projects that capture and store water far from where the water is needed, culminating in the massive water facilities that dominate parts of our landscape."[1]

Gleick acknowledges that the hard path brought great benefits to hundreds of millions of people:

> Thanks to improved sewer systems, cholera, typhoid, and other water-related diseases, once endemic throughout the world, have largely been conquered in the more industrialized nations. Vast cities, incapable of surviving on their local resources, have bloomed in the desert with water brought from hundreds and even thousands of miles away. Food production has kept pace with soaring populations largely because of the expansion of artificial irrigation systems that now produce 40 percent of the world's food. Nearly one-fifth of all of the electricity generated worldwide is produced by turbines spun by the power of falling water.[2]

We need more water than ever, for more and more purposes and for more and more people. Gleick is now a skeptic, but the hard path can perhaps still bring us some of that extra water.

There are three main ways of acquiring water where there isn't enough — where the supply is over-allocated, or where the population has increased heedlessly, or where drought has diminished the supply in the first place.

The first and easiest is to conserve and store water for use in times of shortages; this implies thrift but it also implies dams and reservoirs, as we have already discussed in some detail. To oppose some dams, in some places, is rational; to oppose all dams, in all places, is destructive. Many countries don't store nearly enough water for their security — Pakistan is a prime example — and while some water can be stored by aquifer recharge, reservoirs behind dams are the only realistic way of capturing the natural runoff. Climate change will make this more essential than ever. Those who oppose dams should therefore oppose bad dams and work to make the others as benign as they possibly can be.

The second way, which often intersects with the first, is to import water from afar from those ever-diminishing places where there is a surplus. Chapter 8 looked at some of the pros and cons of mass water movement. The same strictures apply here as to dams: some bulk transfers are good, some very bad. Upstream and midstream water should not be bulk-transferred from one water basin to another, and the fact that it has been done before (southern California) and is being done again (China's North-South Carrier) doesn't make it any better. But there are hundreds of places where bulk transfers are sensible and doable and should be encouraged. We should not laugh at dreamers like James Cran, with his Medusa bags: taking water from the Columbia estuary to San Diego and points south is neither destructive nor impractical. Yes, San Diego should conserve more and desalinate more, but these efforts can easily coexist with shipping water in bulk.

The third way of adding to the usable water stock is to manufacture it from water that is not otherwise usable — from brackish and contaminated supplies, even from fracking's wastewater, or from the oceans. This means stripping from the water the stuff that makes it undrinkable, dumping that stuff somewhere, and injecting the clean water into the distribution grid where there is one. This is generally called "desalination," though it is not always ocean water that is the source or salt that is being removed — the word has become a generic term for converting any non-drinkable water into water that's safe. We'll explore this solution in some detail below.

All three solutions are capital-intensive endeavours, which is why they have been dominated by the wealthy countries and transnational corporations, and why private capital has been enlisted in poor countries or in regions otherwise strapped for cash. As discussed in Chapter 5, rote opposition to private involvement in water delivery can hamper and prevent efforts to bring safe water to the world's poor; the social justice movement should continue to oppose profiteering and exploitation but should adjust its efforts to make sure privatization of utilities is done in the best way possible, with the best possible contracts and comprehensive safeguards.

All three of these approaches have historically been resoundingly successful because of the economies of scale so familiar from mass manufacturing. A modern economy cannot function if half its citizens must fetch water from distant wells in buckets, taking most of an unproductive day to do so. It is much more efficient, and cheaper in the long run, to build and operate a communal water distribution system through pipes.

In many ways, all three of these system improvements still work and still deliver economies of scale. But as it becomes more difficult to expand existing networks to serve new populations or industries, whether because of the expensive extra energy required or the need to find fresh supplies somewhere else, or through increasing leakiness and waste, the cost per unit of water begins to rise. At some point, as Peter Gleick and many others have often pointed out, once economies of scale have been exhausted, the marginal cost (the cost of each additional litre) of water from piped systems will sooner or later become higher than the marginal cost of water conservation efforts. In many parts of the world, we have already passed that point. That will be the subject of the next chapter.

So let's consider the third way: make more water as a way of reaching water security. The most obvious method is to clean up water that is dirty, or saline.

How Big Is the Desalination Industry?

Desalination has become a popular backup technology for many communities, not just in the American West but globally. Perth, Australia's driest big city, would not have got through the Big Dry drought without it. In other places, desal is more than a backup: it is becoming a primary source of water.

There are plenty of reservations about desalination as a solution to water problems and we will come to these. But first, we need some sense of the scale of desal and what its proponents say about it.

As of 2013, the International Desalination Association (IDA) listed 17,277 "water factories" around the world, up from 4,000 a decade before, with a total capacity to produce 80.9 million cubic metres of potable water every day. To try to put this number into perspective, the IDA suggests that this would be about the same as thirty-two years of rain falling on Londoners' heads, or twenty-one years of rain for New York, a more or less useless comparative. Is 80 million cubic metres a day a lot? Compared to what? If you multiply it out annually, it means desal is producing just shy of 30 billion cubic metres a year — but the global demand for fresh water is *increasing* by 64 billion a year. So by this measure, desal is not going to solve a global problem by itself.[3]

This doesn't mean desalinating water is useless; far from it. These are just global averages and tell you little about the usefulness of desal or how it has brought water security to places where there was none before, or how it is being used as a bridge technology along the way to something better. Like other solutions to water problems, desalination is a local issue.

To put it in a single country perspective, Israel currently produces 1,532,723 cubic metres of desalinated water a day — about 40 percent of the country's freshwater needs. That's more than a drop in its particular bucket.

Israel has been building these plants at a breakneck pace, reflective of the urgency of its need. The country's first plant opened at Ashkelon only in 2005, but more have been added since, at Ashdod, Hadera, and other places, with others on the drawing boards. The newest, the Sorek desal plant on the Mediterranean coast fifteen kilometres south of Tel Aviv is the largest and most advanced and it alone produces somewhere around 20 percent of the country's municipal water, up to 624,000 cubic metres a day at full capacity. The costs are pretty low too — the plant can provide all the water for an average family for between $300 and $500 a year. Avshalom Felber, boss of the company that built and operates Sorek, IDE Technologies, points to the other advantage: "Basically this desalination, as a drought-proof solution, has proven itself for Israel. Israel has become ... water independent, let's say, since it launched this program of desalination plants. By meeting its water needs, Israel can [now] focus on longer-term agricultural, industrial and urban planning."[4] Israeli academics have also suggested that the country can start using its expertise to solve regional water problems. As Jack Gilron of Ben-Gurion University put it, "In the end, with everybody having enough water, we take away one unnecessary reason that there should be conflict."

A 2014 survey of the issue by the Israeli newspaper *Haaretz* begins this way: "After experiencing the driest winter on record, Israel is responding as never before — *by doing nothing*. While previous droughts have been accompanied by impassioned public service advertisements to conserve, this time it has been greeted with a shrug — thanks in large part to an aggressive desalination program that has transformed this perennially parched land into perhaps the most well-hydrated country in the region."[5] It is also something of an exaggeration, as we saw in the water conflicts chapter, to say that the region can now coast along with a modest surplus. This understates the region's problems by some orders of magnitude. And relying on one or two desal plants for so much critical water leaves the supply open to sabotage and military attack, something the Israelis know only too well.

Still, for the moment, the supply seems secure. In the winter of 2014 it was so secure that the national water company Mekorot actually reduced its demand from the desal plants to about 70 percent of production capacity. The other 30 percent is a sort of insurance policy against future drought and future demand. Israel is "no longer dependent on the mercy of God to give us rain," as a scientist for the Ministry of Energy and Water Resources, Shlomo Wald, put it.

There are geopolitical advantages too. Israel is a small country, but its scientists have advised China, the United States, and other countries on building up their own capacity. IDE Technologies, the company that built Sorek, has been hired to construct a huge desal plant at Carlsbad, in California, and has already constructed China's largest desal plant. As Wald said, "Israel is the heart of know-how in desalination worldwide. We don't manufacture the membranes, we don't manufacture the pumps. But the engineering and the way a desalination plant should be designed and built, I think, the international hub is here in Israel."[6]

It is no surprise, then, that the Israeli embassy in India sponsored a trip for Indian water experts to visit desal and wastewater plants in Israel and to "sit down with their Israeli counterparts to hear about the newest technologies that help keep Israel green." One of the visitors was Rajeev Jain, water expert in Rajasthan, who made it clear why he was there: "In India, we have a major crisis of water. Our problem is the same that Israel faced, but Israel is an expert at successfully implementing technologies that we aren't able to implement. So we have come here to understand which technologies they use and how they manage these things." Debra Kamin, a features writer for the *Times of Israel*, talked to some of the visitors and found they were uniformly shocked at how expensive water was in Israel, and even more shocked to discover that all citizens, regardless of income or place of residence, must pay the same amount. That wouldn't be possible in India, Rajeev Jain told her, somewhat ruefully, "In India, much of the water generated by cities is illegally siphoned off by residents or lost to leaks, and in rural areas, most farmers get their water at no cost. [We] consider water a gift from God. And everything God has given, no one can charge for it," he said. And he added, somewhat redundantly, "It is not easy to frame new policies."[7]

Some of the Indian delegates to Israel had also been to Singapore, another world locus of desal expertise. Singapore is a country with virtually no water resources of its own (except rainwater collection) and is famously a master of water management. Late in 2013, Singapore officially opened its second desal plant, a 124-hectare facility called Tuaspring that can produce 265 million litres of potable water daily, tripling the amount it already gets from its earlier desal plant, SingSpring. This would account for about a quarter of the country's daily needs, providing water at around 45 cents a cubic metre. Cutting the ceremonial ribbon, the prime minister, Lee Hsien Loong, took the opportunity to pat his countrymen on the back, reminding them of what it had been like when independence from Britain finally came in 1965: "We were almost totally dependent on water supply from Johor [Malaysia]. Singaporeans lined up at public taps for water, and employed night-soil collectors because homes lacked sanitation. But the republic has since turned a strategic weakness into a source of thought leadership and competitive advantage." Political leadership made all the difference, he added virtuously, and largely accurately. Certainly, without the vision of his predecessor (and father), Lee Kuan Yew, nothing like it would have been achieved. The earlier Lee pushed through the notion of self-sufficiency, enlarging Singapore's water catchments, upgrading infrastructure, building a deep sewerage system, and hiring private companies to build desal capacity. "The Government also engaged industry in public-private partnerships to explore and pilot new technologies and develop water infrastructure," as the current Lee put it.[8]

China, for its part, might be planning massive water diversions, but it is not neglecting desal either. And, typically, the Chinese are thinking big. By 2019, a new desal plant in Tangshan, in Hebei Province near Beijing, should be producing a prodigious million cubic metres of fresh water daily, as much as a third of the needs of Beijing itself, a city of twenty-two million people. This is an add-on second phase of a project being built by Aqbewg, a joint venture between Aqualyng, a Norwegian company, and Beijing Enterprises Water Group, headquartered in Hong Kong. Phase One, using water drawn from the polluted Yellow Sea, produces about fifty thousand cubic metres of water each day for the district's use. The project's developers estimate the cost to consumers at $1.29 a tonne, double the current price but cheap at the price of the alternative, which is no water at all.[9]

So far there are few desal plants of any consequence in the United States, but this is changing. The largest plant in the country is at Tampa Bay, Florida. Another is at El Paso, Texas, and holds the record for being the furthest desal plant from the sea (El Paso cleans wastewater, not seawater). California, stricken by drought and planning for conservation, is not neglecting desal either. The Israeli-built billion-dollar

Carlsbad plant, supposed to be ready by 2016, will be the largest facility of its kind in the western hemisphere, producing around 190 million litres of drinkable water every day. This is only one of the seventeen desal projects planned for the California coast as of 2014. The little resort town of Cambria, not far from William Randolph Hearst's castle at San Simeon, is one of them; the town hurried a desal plant into production in less than a year. It will also recycle sewage wastewater, combining treated sewage with estuary water and groundwater to provide a third of the town's needs.[10]

Carlsbad, big as it is, is not going to solve California's problems or substitute for water sourced elsewhere. Just to compensate for the water imported from other river basins, southern California would have to build a Carlsbad-scaled plant every five kilometres along the coast (twenty-five plants between San Diego and Los Angeles alone), solving only one aspect of the problem by industrializing the entire Pacific coast, a prospect impossible to contemplate politically, and in any case anathema to the West Coast ethos, never mind to the surfer-dude and the rich-mansions-on-the-sea voting blocks. This is not to say some of them shouldn't be built. This is not an either-or decision: in a quiverful of solutions, desal is but one arrow. The general manager of the metropolitan water district of southern California, Jeffrey Kightlinger, admits desal is needed and helpful. "There are two things that are changing the landscape for us," he says. "One is we've grown a lot. We're doing water for nearly 40 million people statewide. The second thing that really changed is the climate. Climate change is real. And it's stressing our system in new ways.... We don't have time to rehash the same debates over and over and over again. We're going to have to start investing in things for the future."[11] Desal is among those things.

How Desal Works

A number of techniques exist for cleaning sea or brackish water, but only one, reverse osmosis (RO), is cheap enough to be used on a large scale.

A technique called electrodialysis, in which salts are removed by an ion-exchange membrane, is in some ways more efficient than RO (it leaves a smaller residue and is thus good for wastewater treatment), but it is more intricate to operate and is difficult to scale up to industrial standards. GE produces a commercial model, used mostly in remote locations, but its market is fairly small.

Distillation, a method familiar to any student who in a high-school chemistry lab produced distilled water for experiments, is simple enough to do: the water is heated to steam to separate out dissolved solids, then cooled back to water in a separate vessel. This is the same method used to produce cognac — or moonshine. Distillation has some advantages over RO: it has no membranes that need cleaning, and the water doesn't need to be pretreated because there are no filters prone to clogging. It also leaves less residue than RO. It is, however, more energy-intensive and therefore more expensive.

Reverse osmosis is simple, if rather difficult to keep operating because of the clogging problem. RO involves pushing dirty water (sometimes pre-filtered to get rid of larger objects such as gravel, weeds, or fish) through permeable membranes designed to let the water molecules through but to keep out dissolved salts, pesticides, and bacteria, whose molecules are larger. Brackish water is usually easier to clean than sea water because it is not as salty. The volume and the quality of the water produced depends on the pressure applied to force the water through the membranes, and on how efficient the membranes are — since they are the heart of the process, their actual design is generally regarded as an industrial secret.

Both RO and distillation produce somewhere between 30 and 40 percent clean water, with the rest being brine's minerals. The resulting water is very clean: distillation leaves less than fifty parts per million of dissolved salts (drinking-water standards are usually around five hundred parts per million); RO is not quite that clean but still is well below levels that would cause concern. Nevertheless, for safety's sake, wastewater desal plants usually add chlorine before sending the water off for consumption.

Downsides of Desal

Residual brine is a major issue of desal. As reported above, for every hundred cubic metres of fresh water produced, somewhere around two hundred cubic metres of highly saline brine is left behind and has to be disposed of. Mostly, this is just dumped into the sea through outfall pipes; since it is saltier and denser than regular seawater, it generally sinks to the seabed, where it can cause problems for local plant life and sea creatures. Most plants therefore push the outfall pipes deep into the sea, preferably where a current is running, which will dilute it faster. IDE's Felber dismisses the impact of returning salty water to the salt sea as minor, but no studies have been done on the impact of multiple plants in a contained sea, and Israel is not alone in building desal plants on the Mediterranean coast — Spain, France, Cyprus, Lebanon, and Egypt have already built some and are contemplating more. So far, France is the only jurisdiction that controls brine outflows; French regulations forbid brine discharge greater than 10 percent above ocean salinity. Most brine emissions are higher than that, usually 1.5 to 1.8 times as saline as seawater.

A novel solution is to mix the brine with cement to produce so-called saltcrete or stonecrete, which started as a way of immobilizing hazardous waste but is now also commonly used for paving walkways, driveways, and, when mixed with asphalt, even roads.

Curiously, desalinated water can actually be *too* clean. Water is one of nature's natural solvents, and water without its mineral buffers can be, as a WHO report put it, "aggressive to cementitious and metallic materials used in storage, distribution or plumbing, and requires conditioning to address this problem."[12] In other words, water that is too pure can dissolve concrete and corrode metals, and you have to "de-clean" it before it can be safely injected into a distribution grid unless it is promptly mixed with non-desalinated fresh water. Generally, the WHO report suggests that somewhere between 1 percent and 10 percent of other water should be mixed into the desalinated reservoirs to make them non-corrosive. Sources are not always easy to find — if clean water is already available, why desalinate dirty water in the first place? Generally, partially treated but still saline sea water is added or, where available, untreated groundwater. For obvious reasons, the report adds hastily if redundantly, "This potential short-circuiting of the main treatment process should not allow pathogens and other undesirable microorganisms to be introduced into the finished desalinated water." In some cases, the desalinated water is not diluted, but corrosion-inhibiting chemicals are added directly to the finished product. These include silicates (sands, mostly) but also orthophosphates and polyphosphates, chemicals already widely used in the world and thought to be benign to human health.

Another potential reservation about desalinated water was raised by a researcher affiliated with the WHO, Frantisek Kozisek, in a paper called "Health Risks from Drinking Demineralized Water,"[13] in which he claims, based on earlier Soviet studies, that water with low concentrations of total dissolved solids causes minerals to leach from the body and can cause intestinal and urinary tract disorders. The most obvious effect of this is said to be diarrhea. Still, even while noting this study in a footnote, the WHO admits that it "remains controversial in many quarters," and other studies, such as one from the Canadian Water Quality Association, have dismissed it as misleading.[14] In a useful study, the US Navy, which has been offering desal water produced by onboard nuclear reactors to its sailors for more than forty years, has found no ill-effects whatever. Still, the WHO does suggest that in regions where populations had become accustomed to natural water with high mineral content, it might be prudent to add calibrated amounts of those minerals into desalinated drinking water.

Desal's Energy Use

Apart from brine, energy use is desalination's major drawback, and as a consequence, many scientists are investigating ways of making desal less energy-intensive. To take but one example of high energy use, Israel's desal water is essential, but it may not be very sustainable — the desal plants currently operating use as much as a tenth of the country's electricity production — and that comes at its own cost. Solving one problem (available water) by exaggerating another (greenhouse gases) is not anyone's idea of ecological balance. Still, energy efficiency has been getting steadily better over the last decade. Peter Gleick's Pacific Institute calculates that, in the United States, desalination plants use around fifteen thousand kilowatt hours of electricity to produce 3.8 million litres of fresh water. This doesn't seem so high considering that it takes about the same fifteen thousand kilowatt hours to import the same amount of water from elsewhere (if an elsewhere can be found). Wastewater cleaning and reuse is more energy efficient than this, using only eighty-three hundred kilowatt hours for the same amount of product, but still, using even less energy would make the whole thing more palatable.

At the start of this chapter, I reminded readers of the fantasy common in the Nixon era that too-cheap-to-meter nukes could produce as much water as we needed. Nuclear energy's proponents — and there are still some — keep timidly pointing out that nuclear energy is, by climate-change standards, a clean and green energy: nuclear power produces virtually no greenhouse gases. It might surprise you to learn that nuclear desalination has been running quietly, and generally successfully, in a handful of countries, sometimes for decades. True, one of them is Kazakhstan, not usually a go-to destination for technological sophistication or for ecological probity. Nevertheless, a fast-reactor at Aktau in Kazakhstan has reliably produced 135 megawatts of power while also desalinating 80,000 cubic metres a day of brackish water for thirty-three years and counting. The plant actually produces 120,000 cubic metres a day, the balance coming from an oil and gas co-generation facility. Japan, before it was spooked by Fukushima, was operating ten desal facilities with reactors. None of them were used for potable water — the output was generally used to cool the reactors — but they could have been. They ran for one hundred reactor-years without incident. The same thing has been done in several places in India, again mostly to provide clean water to cool reactors, but also to produce potable water; Pakistan has done the same thing at its Karachi Nuclear Power Complex, and so has China.

If you're going to use conventional power sources, though, it be-hooves you to use the energy as efficiently as possible, so it is essential to improve the efficiency of the RO process. Here, lots of science is being done, if so far none of it advanced enough to take to market. An example is a process invented by a chemical engineer from the New Jersey Institute of Technology, Kamalesh Sirkar, who has devised what he calls a "direct contact membrane" that can extract fresh water from water that is 20 percent brine (typical seawater is 3.5 percent salt). This means he can process seawater several times, thereby increasing the yield. He figures he can produce about eighty litres of drinking water from one hundred litres of brine, a substantial increase.[15]

One of the most promising leads is a collaboration between Lockheed Martin, the aerospace company, and MIT; it involves using graphene, an allotrope of carbon that rather resembles, as the *Economist* once put it, "An atomic scaled chicken wire." Most of the research into graphene has so far been focused on electricity generation (graphene is the best conductor of heat yet found), but membranes made of the sub-stance contain nano-sized holes perfectly sized to let water molecules through yet still hold back hydrated chloride and sodium ions. As a consequence, far less pressure is needed to force the water through its filtering membrane, reducing energy costs by a startling two-thirds.[16]

Even without graphene, modern membranes are twenty times more efficient and one-fifth the cost of the earliest versions tested sixty years ago. And Israel's IDE is generating power by using high-pressure brine from the desal plants to help rotate the pump motors that force the water through the membranes in the first place, a process it calls energy recovery. A standard turbine can recover about 80 percent of the energy input; IDE's process takes this to 96 percent.[17] Using this technique and other innovations, IDE believes that its Carlsbad plant in California will actually become carbon-neutral and, in addition, will use $12 million less fuel to operate each year.

What about using the sun or the wind to desalinate water? A test solar desalination project was running in 2014 in California's parched Central Valley, in this case not to desalinate seawater but the billions of tonnes of contaminated water that lies beneath the surface, water so charged with toxic levels of selenium and other heavy metals that it must be periodically drained away to avoid poisoning crops. The company running the test, WaterFX, is effusive about the possibilities. As the company's website puts it, "Renewable desalination is a global solution with the power to create freshwater from abundant resources such as solar energy and saltwater.... [It is] the kind of solution that is market driven, not government subsidized; it is politically, legally, financially, and sustainably achievable in the near term and is an approach that can put arid regions on a path towards true sustainable water independence. The data being generated by the... pilot proves it is a viable alternative to generate new sources of water alongside conservation and reuse."

Instead of using reverse osmosis, WaterFX uses an off-the-shelf four-hundred-kilowatt parabolic solar trough designed for power generation to concentrate the sun to heat a reservoir of liquid that then transfers to a heat pump to boost output, all of which is then used to evaporate water and condense it out as pure H_2O.

The test project, paid for by the Panoche water district, can produce 1,233 cubic metres for $450, twice what farmers used to pay the Central Valley Project when water was available but less than they were paying late in 2014, when prices ranged from $500 to $2,200. When the actual plant replaces the test project, it will cover thirteen hectares of land and produce about a quarter of a million cubic metres.

Another promising project was unveiled by IBM in 2014. The firm's nine-metre-tall "sunflower" was designed for solar electricity generation, but it has an intriguing side product: very hot water that can be used to desalinate unclean water. The sunflowers — so named because they do rather resemble giant metallic flowers, and they do rotate to follow the sun — can convert, or so it is claimed, 80 percent of the sun's energy into water and power more efficiently than conventional systems and a good deal more cheaply. The key development is to use microscopic tubes that carry water through the cluster of photovoltaic chips the device comprises — the same system IBM uses to cool its supercomputers, inspired, as IBM engineer Bruno Michel put it, "by the branched blood supply of the human body." One such sunflower could provide power and water for several homes; a "plantation" of them could easily provide enough clean water for a small town, the company says.[18]

The Sahara Forest Project

A larger-scale experiment is under way in Jordan, at Aqaba, where the Red-Dead project starts. The project uses a basket of conventional technologies, but in novel ways, to make fresh water. If they get it right, it will produce not just water but also food and electricity, which means that it has the potential to green large swathes of coastal desert. It is the brainchild of an environmental technology group based in Norway, a long way from the nearest desert. Still, governments are taking it seriously: in January 2011, Norway and Jordan signed a formal agreement to get the project underway, with funding finally approved in 2014. The twenty-hectare demonstration site is scheduled to open sometime in 2015. Four hectares will be greenhouses, the other sixteen solar reflectors, support buildings, and open-air crops. As its proponents put it, "The system is designed to utilize what we have enough of, to produce what we need more of — using desert, saltwater and carbon dioxide to produce food, fresh water and energy."[19]

The Sahara Forest Project (SFP) works by bringing seawater into the desert and evaporating it. The SFP website explains it this way: "Sahara Forest Project combines solar thermal technologies with technologies for saltwater evaporation, condensation of freshwater and modern production of food and biomass without displacing existing agriculture or natural vegetation; a single SFP-facility with 50 MW of concentrated solar power and 50 hectares of seawater greenhouses would annually produce 34,000 tons of vegetables, employ over 800 people, export 155 GWh of electricity and sequester more than 8,250 tons of CO2."

According to the company's website, the three core components are:

- Saltwater-cooled greenhouses — greenhouses that use salt-water to provide suitable growing conditions that enable year-round cultivation of high-value vegetable crops even in desert conditions. By using seawater to provide evaporative cooling and humidification, the crops' water requirements are minimized and yields maximized with a minimal carbon footprint.
- CSP (concentrated solar power) for electricity and heat generation — the use of mirrors to concentrate the sun's energy to produce heat that is used to make steam to drive a steam turbine that powers a generator to produce electricity. The solar plant reflectors will concentrate sunlight onto a pipe carrying a heat-absorbing fluid that produces the steam.
- Technologies for desert revegetation — a collection of practices and technologies for establishing outside vegetation in arid environments, such as evaporative hedges.

Daniel Clery, in a piece for *Science*, says the key is an earlier version of the scheme, the "seawater greenhouse," developed by British inventor Charlie Paton, who has already built demonstration projects in Tenerife, Abu Dhabi, and Oman.

> In Paton's scheme, seawater piped to the greenhouse trickles down over a grid structure that covers the windward side of the greenhouse. As natural breezes blow into the greenhouse through the grid, it evaporates the water, making an interior that is cool and moist — ideal conditions for growing crops. At the other end of the greenhouse, another grid evaporator, fed by seawater heated in black pipes on the greenhouse roof, loads more moisture into the air as it leaves the growing area. Now hot and very humid, the air passes through a maze of vertical polyethylene pipes cooled by cold seawater passing through them. Fresh water

condenses on the pipes and trickles down into collectors,
to be used for irrigation or drinking.[20]

Curiously, the SFP designers used as one of their models a peculiar
Namibian beetle that has evolved a ribbed fantail, on which fog condenses, then drips down the ribs and into a small sac. Israeli scientists
have already explored and mimicked this ingenious invention — an
experimental station in the Negev uses fog-drip irrigation for small-
scale agriculture, creating cones of moisture-trapping fibres around
individual plants.

Jordan has long been exploring the notion of building such seawater
greenhouse plants alongside its Red-Dead pipeline.

A pilot plant built by SFP in Qatar produced seventy-five kilograms
of vegetables per square metre in three crops annually, a comparable
value to commercial farms in more-verdant Europe, while "consuming" only sunlight and seawater. The Qatari plant uses the technology
described above: mirrors in the shape of a parabolic trough heat a
fluid flowing through a pipe that then boils water, the steam driving
a turbine to generate power. Hence, the plant has electricity to run its
control systems and pumps and can use any excess to desalinate water
for irrigating plants. By one estimate (from the designers and yet to be
demonstrated), a sixty-hectare greenhouse field "could provide all the
cucumbers, tomatoes, peppers, and eggplant now imported into Qatar."

On that note, we should leave the last word to Ezekiel, the prophet.
Here he is on foreseeable consequences, with a touch of sunny optimism:

On one side and on the other, will grow all kinds of trees for
food. Their leaves will not wither and their fruit will not
fail. They will bear every month because their water flows
from the sanctuary, and their fruit will be for food and their
leaves for healing.

15
Soft-Path Solutions

The key concept is simple: don't increase the water supply unless you really have to. Increase water efficiency instead.

The term "soft path" seems to have been coined by Amory Lovins of the Rocky Mountain Institute think tank. He was talking about energy, not water, but his soft path was nevertheless conservationist in essence: instead of steadily increasing the supply, steadily decrease the demand. One of his aphorisms was the "negawatt," that is, the watt that didn't have to be generated because there was no longer need. Peter Gleick's Pacific Institute adopted the "soft path" phrase to describe water-management issues and it has since become common currency in the water world.

Gleick puts it this way: "We have reached a fork in the road. We must now make a choice about which water path to take. We know where the hard path leads — to a diminished natural world, concentrated decision making, and higher economic costs. The soft path leads to more productive use of water, transparent and open decision making, and acceptance of the ecological values of water."[1] I think this overstates the case, but in essence he is right.

In 2003, the International Rivers network, an often-sensible NGO with much good work to its credit, published a paper called "Crisis of Mismanagement: Real Solutions to the World's Water Problems" with the main intention of countering the World Bank's famous "dismal arithmetic of water" concept: The paper opens this way:

We are widely perceived to be in the midst of a "world water crisis." This crisis is commonly believed to be one of scarcity — that the world is running out of water. But in fact, the "crisis" is mainly one of mismanagement, not absolute scarcity. Freshwater ecosystems worldwide have been dammed, drained and pumped dry to supply inefficient and inequitable irrigation schemes, leaky water mains and wasteful overconsumption....

More than a billion people lack access to decent water supplies, not because there is too little water, but because governments have failed to provide it. Just one percent of current water withdrawals would supply a basic level of 40 liters per capita per day to all those currently lacking adequate supplies — and to the two billion people projected to be added to the world's population by 2025.

The paper then goes on to list a number of soft-path initiatives, all of them sensible and eminently achievable; they overlap with my own list that is itself a gloss on Gleick and other water managers. There are a few flaws — the Rivers people can't resist yet another perfunctory slap at dams and privatization, and that "just one percent of current water withdrawals" is disingenuous — as I said earlier, you can't reduce withdrawals in one part of the world and hope for it to have an effect a continent away. The assertion is very much akin to that of the renewable power enthusiasts who talk about capturing "just one percent of the energy from the sun that falls on the earth every day," as though that were technically feasible and was just waiting for a policy directive from on high.[2]

If the soft path means anything at all, it means demand management. It means that if you can be convinced that you need less water, then no one will be obligated to supply what you no longer need. In this, inertia is the enemy, and habit — not changing because it is easier to just keep on keeping on, using what was used before without any care for thrift.

It's that word "need," though. It hides a multiplicity of complications. It's very easy to think of your wants as your needs.

Soft-Path Strategy 1: Give Water Value and Price It Properly

Almost everywhere water is far too cheap, and in far too many places individuals, corporations, and municipalities are allowed to extract as much as they want without any regard for the real cost of providing it — or for the effect withdrawals have on the environment. In a survey of fifty Asian cities, for example, only half of them even attempted to measure water allocations to their citizens, who therefore had no incentive to conserve.

This is one of the negative consequences of recognizing that water is not a commodity like any other, that it is, unlike oil, essential to life. If it *were* treated as a commodity, there would have to be some relationship between consumption and the cost of delivering the water in the first place. After all, in elementary economic theory, in any given consumptive use, the price equates with the marginal cost of providing the good involved — factoring in scarcity, the cost of replacing the good with alternatives or other supply sources and the ecological consequences of doing so. Instead, what we have are price caps, untenable subsidies, and an historic refusal to even consider amending regulations that date from days of plentiful supply and thin populations. Water subsidies began during the Industrial Revolution in Europe, and in many countries have hardly changed in the interim.

Put the price of water up, usage goes down — this venerable economic truism has been amply proven, in many jurisdictions. Towns that installed meters instead of a flat price saw consumption go down by as much as 30 percent. And yet, even now, in a time of scarcity, not every utility follows suit. By 2014, the city of Sacramento, to take one example of many, had installed meters on fewer than half the homes in the city; the rest could leave their taps running all day and all night without penalty.

One of the concerns about assigning water a value — at whatever level — has been that it would penalize the poor, who need water just as much as the rich but by definition haven't the money to pay for it. As discussed in Chapter 5, this not an either-or issue — there doesn't need to be one price for all. Nor do there have to be means tests to assign basic water to the poor. Even from a social-democrat point of view, increasing the price of water is the right policy — everyone needs a basic allocation, yes, but it is important to price water so that waste hurts. The political trick is to balance conservation incentives against the irreducible needs of those with little money. One of the first to recognize this was post-apartheid South Africa, whose water minister, Ronnie Kasrils, deftly manoeuvred between the twin needs of basic supply and cost recovery. As he put it, it is entirely possible to harmonize the social and economic values for water: "In South Africa, we treat water as both a social and an economic good. Once the social needs have been met, we manage water as an economic good, as is appropriate for a scarce natural resource. Some non-governmental organizations and international organized labor oppose what they call the commodification of water, and thus oppose [any kind of] cost recovery." This is wrong-headed, he asserted. "Absence of cost recovery leads to inadequate funding for infrastructure development, and the resulting overuse leads to local shortages and service breakdowns which impact most heavily on the poor."[3]

The key phrase in all this is "once the social needs have been met." And to some extent, Kasrils believes his government didn't go nearly far enough in providing for the poor — the ANC government made a Faustian bargain with the international development agencies after independence, he now says, and threw away the chance for a more radical economic transformation.

Nevertheless, under his watch, South Africa introduced a tiered-pricing plan for water as well as for sanitation. This means that everyone gets a basic allocation, usually either free or at a heavily subsidized price. All other levels of service — whether a farmer's sprinkler system, a homeowner's Jacuzzi, or a car-maker's need for cooling water in its industrial processes — are subject to escalating fees depending on the amount of water used.

There were hiccups along the way. How to measure water use from communal standpipes where home delivery wasn't possible? How to make sure that the poor actually got what they were allocated? Remember, from the earlier chapter, that the government flubbed its first attempt, issuing prepaid water "credit cards" that the poor couldn't afford, leading to angry protests and endemic water theft. But slowly, more sophisticated water-metering devices were introduced and the hubbub has been dying down.

Many other utilities in other jurisdictions have been following suit, from Osaka, in Japan, to a slew of Californian cities, including Los Angeles. The Indian community of Sawda Ghevra, on the northwestern outskirts of Delhi, has experimented with a decentralized water system using the equivalent of bank ATMs to try to ensure that residents had access to clean, year-round water. Using their own canisters, users could fill up as much as they liked for somewhere around two cents per litre, using prepaid cards. It was a lot cheaper than the water hitherto provided by water tankers, and the purification plant that provided the water is powered by the sun.

While raising prices is a better strategy than the nanny-ish notion of, say, making watering your lawn a crime or insisting that you water only at night or that you rip up your lawn for some trendy version of xeriscaping, the price signals have to be meaningful. Most Americans, after all, still pay only a dollar or so a day for water. In Fresno, California, for instance, a family that uses 1,500 litres a day gets a monthly water bill of $28.26, and doubling consumption only doubles the cost, from trivial to still pretty trivial. But even if the cost does go up, how much is enough? In New Mexico, prices increase "dramatically" — which means fourfold — if a family uses more than 26,500 litres a month. But that fourfold goes from a mere $10 to only $40. Will a family earning better than $100,000 a year balk at paying an extra $30 a month if that's what it costs to water the lawn? And if the levels are punitive, are they really aimed at reducing consumption or merely at matching costs to revenue? Will the prices take into account the long-term ecological consequences of continuing drought? Or of taking water from rivers and aquifers that are not sufficiently replenished? Will they take into account who, or what entity, "owns" the water or has the right to use it? Would freer water trading help, the way it did in the great Australian drought, by ensuring that water goes where it can do the most good? These are political decisions, and they are nothing if not complicated.

A pristine example of how to get pricing wrong was offered up early in 2015 by the government of British Columbia, an otherwise ecologically sensible jurisdiction. In tabling the so-called Water Sustainability Act, much fanfare was made of the notion that, for the first time, the province was regulating groundwater and making it subject to a list of fees and "rentals." But the prices! A typical household would pay somewhere around $1 to $2 more a year, and industrial users, which hitherto had paid nothing, would now be charged the not very impressive figure of $2.25 per million litres. Nestlé, Canada's largest bottled-water producer, would be charged at the "highest industrial rate," according to the environment ministry. What would that be? The 265 million litres the company draws for its plant in the town of Hope would set the bottom line back by $596.25 a year. The environment minister, Mary Polak, said virtuously that British Columbia wasn't selling the water, just charging an administrative fee. "We're quite proud," she said in the legislature, "that BC has never engaged in the selling of water as a commodity." [4] *Tant pis.*

Political decisions, once taken, can also be overruled by an irate citizenry, as Ireland's government found to its cost in 2014. Ireland is not exactly short of water — the stuff flows from the sky more often than would seem polite — but the government nevertheless looked to water for a new source of revenue, exactly the wrong motivation for pricing it. In any event, the announcement that Irish households would henceforth have to pay for the water they had been used to getting free set off a massive backlash, with more than a hundred thousand people taking to the streets, "an astonishing number," the *Economist* noted, "for a country of 4.6 million people with no tradition of mass protest."[5] Defenders of the government's policy noted that Irish citizens use far more water per capita than those European countries that do meter water use. But less than a week after the protests, the Irish government announced that the charges would be capped and would not penalize heavy use. After "conservation grants," the caps would mean single-adult households would pay 60 euros and other households 160 euros. The environment minister, Alan Kelly, said rather ruefully that "anger is never a good starting point for a key decision," and that "the absolute maximum cost [will be] just over three euros a week — much less than one percent of most people's incomes or benefits and puts water bills among the lowest in Europe."[6] The end result is a policy that does bring in some revenue, though not very much, and that will have virtually no impact on usage.

Pricing for farmers adds another level of complication to setting water policy. In California's Central Valley, farmers used to pay about $20 per 1,200 cubic metre for water, while cities paid $200. But by 2014, prices had already escalated far beyond those numbers. Those prices, farmers say, will mean taking some farms out of production, adding to food-chain miseries, and creating shortages. And some well-meaning efforts have had unintended consequences: in the American Southwest, millions of dollars have been poured into helping farmers buy more efficient irrigation systems, only to find that the new systems made the farmers want to irrigate even more acreage, using ever more water.

On another level, utilities (and political authorities) need to re-examine the "right" to extract groundwater or river water at no cost to the user. Even in the US West, counties and states are rethinking the notion that groundwater belongs to the person who owns the land above it. In many cases, states are now requiring permits to extract any water at all. In Chapter 5, on privatization, I described how the Campbell Soup Company in Napoleon, Ohio, withdraws at no cost vast quantities of water from the Maumee River to make its products, and how the company defends itself by pointing out that other users pay nothing for the water they extract from the river either. If the company was charged for that water, it would either pull up stakes and go elsewhere, find more efficient ways of using the water, or pass on the increased costs to the consumer. Of the three strategies, there is not much doubt which one it would choose — it would decamp. So I'm not saying pricing would be easy, or even that it will be universal, only that it is necessary. And if applied, would make a huge difference — measureable, significant, with a real reduction in demand, and therefore less pressure on supply.

Soft-Path Strategy 2: Find Waste and Stop It

The World Bank estimates that leaky infrastructure costs the world's water utilities better than $15 billion a year in revenue foregone — maybe forty-five billion litres a day go missing, or forty-five million cubic metres or, to use the old favourite, eighteen thousand Olympic-sized swimming pools. Losses are thought to average between 20 and 30 percent in developed countries, closer to 50 percent elsewhere. This is almost certainly a conservative number; the real losses are likely to be much higher.

In cities like London, where the water pipes are more than a hundred years old, leaks are commonplace, and this is expensive water that has already been collected, stored, treated, and then delivered, sort of. "Pipe assets," as the water world calls the supply grid, generally last a long time, but long is not forever and in many cities that were early to provide piped water, the systems are reaching the end of their useful lifespan.

The world will likely have to spend somewhere around $23 trillion over the next two decades on water infrastructure, with multiple billions a year to maintain what is newly repaired. The United States alone will have to invest around $45 billion a year; the American Water Works Association expects that repairing and expanding drinking-water systems alone will exceed a trillion dollars.

In developed countries, the price will be paid through higher water bills and local levies. Elsewhere, solutions remain elusive except that they will be tackled piecemeal and that full repairs will take decades.

Until recently, finding leaks in aging municipal systems was a pretty low-tech affair, mostly consisting of waiting until the leaks did actual damage to streets or buildings. Not that low-tech solutions didn't sometimes work when done diligently. On the other hand, a good example of a truly low-tech system that did *not* work was in London, one of the world's great cities with one of the world's great water delivery systems. London's leak-tracking was for decades risibly poor: workers were issued "listening sticks" to push into the ground, put the other end to their ear, and listen for odd noises. Miraculously, they actually found some leaks that way.

Thames Water, the private utility that delivers Londoners' water, failed in its self-set leakage reduction program ten years in a row. Yet, by 2012, it had achieved five consecutive annual targets, by contracting with an Israeli start-up called TaKaDu that uses proprietary algorithms to assemble data from various sensors along the network (flow meters and pressure sensors) to pick up anomalies. If something goes wrong — it could be a small leak, a massive leak, faulty equipment, or just a technician who left a valve open — a signal is sent online to a monitoring station in Tel Aviv. From there, an electronic alarm is triggered and TaKaDu can instantly see the location, the magnitude, and the timing, and send a repair instruction to the utility's management. The interaction takes mere seconds.

TaKaDu, under founder and CEO Amir Peleg (self-described as "chief plumbing officer"), started as a monitor for Israel's water-supply system.[7] But the Internet is no respecter of boundaries and the company soon acquired clients elsewhere, such as Thames Water, also the Yarra Valley Water in Melbourne, Australia, and Aguas Antofagasta, a water utility in Chile, where water losses were reduced from 30 to 23 percent in five years. "Every cubic meter we save means we have one cubic meter less to produce in our desalination plant, which is very intensive in energy," says Marco Kutulas Peet, the Chilean utility's general manager.[8]

Another start-up is a company called Miya, founded by American-Israeli billionaire businesswoman Shari Arison. Miya expanded rapidly after it helped the utility that delivered water to the western side of Manila, in the Philippines, reduce its water losses from 1.5 billion litres a day to 750 million in just a few years. Miya's technicians understand that you can't just plug a leak and be done with it — the whole water system is pressurized, and fixing one leak actually puts added pressure on other parts of the grid — you have to treat the system as a whole. One way of doing that, counterintuitively, is to divide a large grid into smaller zones, and plant multiple sound and pressure sensors in each zone at key points along the grid. Most of their "listening" is done late at night, since most of the water moving through the pipes at night is actually from leakages.

Both TaKaDu and Miya are market leaders, though still small beer. IBM is neither small, nor exactly a start-up. Except in this way: it is applying its data-handling expertise to real-world problems in what the company calls its Smarter Planet initiative. This expertise is applied to everything from smart cars to monitoring medical patients and, in this case, water conservation, or "gaining insights from rain to drain," as a company sales pitch put it. Their argument is that conventional utilities with generally outmoded and isolated IT systems are ill-equipped to deal with exploding data volumes, new data types, and new instrumentation. Done properly, smart water systems don't just find and fix leaks; they can often predict failure. This means that action can be taken to fix pumps before they fail, and to "repair" leaks before they happen. Since the business unit's founding in 2009, IBM has been working with several utilities in the United Kingdom, the United States, Australia, Ireland, and Malta. The same analytics handle freshwater delivery, wastewater disposal, smarter irrigation (mostly for parks and golf courses, so far), and urban flood management. Robert Musgrove, who heads IBM's Global Business Services, suggests that running utilities is only a small part of the ambition. He's going after bigger game — managing whole river basins and estuaries is a part of it. The aim is to build smarter networks to help shape the way people actually use water: "The challenges...are complex, [but] there are solutions. The world is more instrumented and is becoming more interconnected and intelligent; the water sector can take advantage of this. Integrating existing and emerging technologies, and doing so intelligently, could hold the key to a more sustainable future for water around the world."[9]

Soft-Path Strategy 3: Concentrate Instead on Why People Want the Water and What They Are Using It For

The conventional approach has traditionally been to supply the water but not to care what people actually did with it, so water utilities and companies took it as their function simply to supply as much water as they could to match whatever demand they perceived — the more water, the better, in that view. In some cases, they even gave discounted prices for bulk users. With water scarcity, however, end use does become an issue, not just for over-keen regulators but also for communal well-being. Once this conceptual leap has been made — to identify and satisfy actual needs rather than just to sell water — finding efficiencies becomes a central task for a utility, rather than a charge on the bottom line. It also implies a change in design sensibilities for many products. Washing machines, for example, were never conceived of as a way to minimize water use; their purpose was to conserve human labour and free it for other purposes. That is now changing as conservation becomes an increasingly acute issue. Another small example is an add-on to the already efficient low-flow toilet in which a bathroom sink sits atop the toilet's holding tank so water that drains from the sink is available for the next flush.

As Peter Gleick put it in his argument for a soft path to sustainability (from which some of this is sourced),[10] people really don't *need* water except to drink and possibly to cook meals, and for basic sanitation and cleaning. They use water for many other purposes, of course, like cleaning clothes (and themselves) and disposing of waste, but if ways could be found to do these things without using water, or by using very little, and to do them conveniently and hygienically, no one would much care. In this sense, low-flow toilets are only a start. It's true that some of these no-water alternatives would be a harder sell in rich-world countries than elsewhere where some of them are already common-place. For example, Gleick himself points out that toilet flushing is the largest indoor use of water in most Western urbanized countries and suggests that "new technology... can manage human wastes without water.... Electrically mixed, heated, and ventilated composting toilets have no odors or insect problems and produce a finished compost that does not endanger public health. These devices safely and effectively biodegrade human wastes into water, carbon dioxide, and a soil-like residue." The makers of these toilets have latched onto this notion, and an industry website gets positively giddy at the possibilities: "If a community were to embrace the total use of composting toilets and appropriate gray water systems, it would have no sewage charges, sewage pipe installations and maintenance costs. The community would also have greatly reduced water costs. It could also reduce its rubbish collection charges through recycling most vegetable matter, and would be able to produce valuable compost and worm castings for sale or reuse in community and private gardens."[11] All true, but as one who actually owns such a toilet, I can attest that it is not yet quite as convenient as advertised and that the "soil-like residue" remains an issue, keeping you closer than you would like to what you would like to get away from.

Technology might yet come to the rescue here. Retrofit kits already exist to convert septic systems where they exist (mostly in rural or peri-urban areas) into treatment systems instead. And where urban densities don't allow septic systems, it may still be possible to have single-household or apartment-block wastewater treatment systems in-house, so to speak. A paper in the journal *Water Research* reported on a membrane bioreactor in the basement of a four-person house that treated all outgoing water, including sewage, into "de-watered sludge" plus usable water. I say "may be possible" mostly because the system still has kinks, among them residual bad odours, but it is likely those will be eliminated in due course.[12]

Another way of coming at this is by a hybrid system: community sewer and treatment systems built cooperatively and managed by the community itself. These commonly cost only a third of traditional centralized systems and use much less water.

A second part of this strategy is to stop over-cleaning water when such cleaning is not necessary. Clearly, we don't need to use meticulously cleaned and hygienic potable water to flush our toilets or to water lawns, yet we do. We're stuck with it because traditional distribution grids use only one set of pipes in and one out. Some communities have already shifted away from this one-pipe-for-all-purposes mode; Orlando, for example, bedroom community for Walt Disney World, now has two distribution grids — a regular one for potable water and a purple-pipe one for recycled or reused water, called grey water in the trade. No new homes or subdivisions in surrounding Orange County can be built without including this dual water system, and lawn watering, for example, can be done only with grey water. St. Petersburg, also in Florida, has operated a similar system for decades. In water-stressed areas, recycling water in this way is by far the cheapest and easiest way of increasing supply. As Antoine Frérot, former CEO of Veolia Water, said in the *New York Times* a decade ago, all the necessary technologies and processes are known and conveniently available: "They are already in place, and the wastewater is there [just] where we need it, downstream of the cities." A Canadian company, H_2O Innovation, is a leading developer of activated sludge and membrane bioreactor technologies for wastewater treatment, and there are many others. The point is to distinguish between consumptive uses of water, like evaporation and runoff into the ocean, where water cannot be reused, and non-consumptive uses, like showers and runoff from excess irrigation, where the water can indeed be reused. Israel, as so often, is the example to follow: the country has helped to close its water gap by recycling a full 90 percent of its wastewater for agricultural purposes. Uniquely,

Israel has devised a management system that balances demand and available resources among the various sectors: municipal, industrial, and agricultural — a classic soft-path solution.

There are two possible levels of water reuse: separating water into potable and grey water, or doing a three-part division — potable, grey, and black water. In this version, grey water would be that emanating from bathtubs and laundry rooms; black water would mean concentrated sewage. Notions to reuse black water have their detractors — it's hard to overcome the distaste factor in any toilet-to-tap scenario — but both grey and black water systems are working in increasing numbers of places in the world. I have already described the system operating in Windhoek, in Namibia. Singapore does the same thing — about a quarter of the territory's potable water is recycled sewage, or "NewWater" as the Singaporeans have dubbed it, delicately not mentioning its source. Like Windhoek's plant, the water-services company Veolia had a hand in construction. Singapore also directs much of its storm runoff to drinking-water reservoirs with only minimal filtration, a testament to how clean the city otherwise is, and how tough its anti-pollution laws.

In 2008, Orange County in California built a 265-million-litre-a-day wastewater treatment plant that produced "product" its managers say exceeds any drinking-water standards in effect — but no one was drinking it because the state regulators still prohibited such use. Instead, it was injected into the underlying aquifers on which half the county depends. The drought changed all that. In 2013, the Orange County water district announced plans for an expansion of the existing facility, upping its output to 380 million litres a day from 265 million; the water so produced will total about a fifth of the supply for the district's more than two million residents. And all this, as its managers boast, using less than a third of the energy needed to desalinate the same amount of water, and less than half the energy required to bring water in from northern California.

Activist politics, known by its opponents as fear-mongering, sometimes intrudes. In the California community of Redwood City, for example, the local council wanted to emulate Orlando but was thwarted by a citizen initiative that trotted out a doctor and a druggist who both testified about the health risks involved. A compromise of sorts was worked out that allowed some reused water, but only in areas where humans were unlikely to encounter it — leaving out parks and playing fields, for example. Proposals for scrubbing grey water also failed in Tucson, Arizona, and in San Diego — despite San Diego having to import four-fifths of the water it uses. In Tucson's case, a number of politicians played the "I'm not a scientist, but it can't possibly be safe" card; others simply believed there were better ways of achieving the same goal. For example, one state legislator suggested that the city limit its growth instead, a sensible if politically unattainable notion.[13]

Part of this strategy for reducing water use is managing water use outside the home, and part of that is managing water for gardening. Las Vegas started this trend after an earlier drought in 2003 when it offered America's first "turf removal rebate program," which has in the interim spent more than $200 million to remove fifteen million square metres of thirsty grass from residences and businesses, thereby saving thirty-five billion litres of water and dropping the city's consumption by a third (golf courses and the many casino fountains are now obliged to use grey water if they want to water at all). In Los Angeles, lawn rebate programs now save somewhere around 180 million litres a year. Dozens of other cities have followed suit. Some ban lawns altogether, others allow them but don't allow watering, and yet others allow watering only at night and employ inspectors to watch out for water scofflaws, fining them on the spot. Still, at the end of 2014, the state of California had hired only twenty-two people to watch for such misdemeanours, and the revenue from fines was hardly pouring in. In fact, on the flip side, there was some evidence of an illegal "water rush" — poachers siphoning off fresh water from mains and selling it to the highest bidder.

In Long Beach, California, the authorities began offering a "water-waster app" for smart phones, to make snitching on neighbours really easy. As the director of the Nelson Institute for Environmental Studies at the University of Wisconsin-Madison, Paul Robbins, told the *New*

York Times: "The era of the lawn in the West is over."[14] In the summer of 2014, the state declared that wasting water — doing things like hosing down a sidewalk to get rid of debris — would henceforth be a criminal offence.

Elsewhere where matters are not so dire, more creative solutions are being found. For example, Minneapolis, its greatest problems not so much supply but overall water management, has devised a number of schemes to encourage ratepayers to conserve water and to manage rainwater where it falls. Homeowners and small businesses are offered a range of techniques for conservation and runoff management — installing permeable paving instead of asphalt, placing a "stormwater pond" or wetland somewhere on the property, planting a "green roof," creating rain gardens, installing rain barrels — anything to help reduce the amount of water that flows off urban land into waterways during high rainfall events. The incentives have been priced appropriately: if a household adopts all of these strategies, annual real-estate taxes can be reduced to zero — the city still benefits from having to spend less on flood controls and storm sewers.

Sometimes quantifying actual water usage can be difficult. As a case in point, the food conglomerate Nestlé has been extracting water from aquifers in the desert area around Palm Springs and shipping it off in plastic bottles under the brand names Arrowhead and Pure Life. No one knows how much water it's extracting because its wells are on tribal lands (the Morongo Band of Mission Indians) and are exempt from reporting. Nestlé has not been forthcoming, refusing to divulge any numbers. In the way of these things, the company instead issued a platitudinous press release meant to be reassuring but which instead has raised hackles by its transparent insincerity: "We proudly conduct our business in an environmentally responsible manner that focuses on water and energy conservation," the company said. "Our sustainable operations are specifically designed and managed to prevent adverse impacts to local area groundwater resources, particularly in light of California's drought conditions over the past three years." It might, of course, even be true. But a little skepticism is not unjustified.[15]

In July 2014, California strengthened its legal hand (outside tribal areas, that is), imposing fines of up to $500 per day for watering a garden or hosing down the sidewalk, a necessary move, state officials said, to counteract a still-apathetic public. In 2014, California consumed around 76 billion litres a month less than just a year earlier, and 2013 was better than 2012. This sounds like a lot, but it is only a 5 percent reduction, when the governor, Jerry Brown, was insisting that 20 percent was necessary. So paying attention makes a difference, but so far not one that's big enough.

Soft-Path Strategy 4: Decentralize Where Possible

This can work on large scales — statewide initiatives in the American West — or tiny — rainwater harvesting and management at the village level in India. As Peter Gleick wrote, "Decentralized investments are highly reliable when they include adequate investment in human capital, that is, in the people who [actually] use the facilities. And they can be cost-effective when the easiest opportunities for centralized rainwater capture and storage have been exhausted."

Solutions can be as simple as "water harvesting" — encouraging homeowners and smallholders to capture rain and snow, storing it either in the soil, in cisterns, or in ponds. Retrofitting homes for water capture is inexpensive and will benefit communities by minimizing stormwater runoff.

On a large scale, a study by the Natural Resources Defense Council, University of California, Santa Barbara, and the Pacific Institute investigated what water savings could be derived using current technologies in the state and found that improved efficiency, water reuse, and, above all, stormwater capture could together save a total of up to 8.8 billion cubic metres of water a year in urban areas alone, enough water to supply all of urban southern California, with enough left over to recharge aquifers. The study also pointed out that urban areas use a fifth of all California's water supply, much of it brought in from reservoirs hundreds of kilometres away at great cost — decentralizing would therefore also save large amounts of energy and money.[16]

The same study estimated that farming, which uses 80 percent of the state's supply, could drop its consumption by up to 8.1 billion cubic metres through improved irrigation techniques. "In total," the study concludes, "these 21st century water supply solutions can offer up to 14 million acre feet [17.3 billion cubic metres] in new supplies and demand reductions per year, more water than is used in *all* of California's cities in a year. These savings would provide enough water to serve 20 cities the size of Los Angeles, every year." Some 17.3 cubic metres is the total shortfall expected from California's drought.

On a medium scale, consider the always-drought-prone Brazilian state of Ceará. Water management for the state has been centralized under a state utility, but that is changing in an initiative called "The Ceará allocation project," designed by the Columbia Water Center in New York. Local citizen committees representing residential, industrial, and agricultural users are now making decisions about seasonal storage and use based on information supplied by the utility. As the Columbia Center reports,

> Dramatic fluctuations in the amount of rainfall the region receives from year to year make this decision-making and planning process extremely challenging, even as divergent interests set the stage for potential conflict among different sectors ... [and the allocation project] allows different sectors to cooperate in innovative ways. For example, with a clearer picture of an upcoming drought, farmers could avoid the risk of total crop loss by not planting in a given year, while ceding their water rights to industry for compensation — thereby transforming a situation characterized by uncertainty and potential conflict into a cooperative, mutually beneficial and efficient process to benefit all residents of the state.[17]

Similarly, in the town of Milhã, where most residents were obliged to travel to distant wells for their water, the Columbia Water Center designed a system that incorporated social analysis and community engagement that allowed the town to build cheap and sustainable local infrastructure for household water — the soft path perfectly encapsulated.

On a small scale, villages in India have been encouraged to create their own water board and distribution systems, relying on communal wells and small rainwater reservoirs. Membership in the managing cooperatives is free, though villagers pay for the system's upkeep, usually a few dollars a month. Of course, many are not making much more than that a day, but the mere fact of having reliable water has increased farm efficiency and incomes. Elsewhere in rural India, more and more women are being brought into local management committees called *pani panchayat*, or water councils, which is having a substantial effect on the reliability of supply. A study by India's National Council of Applied Economic Research found that rural women on average spent nearly a fifth of their day fetching and carrying water. Not surprising, then, that where local councils were dominated by men, they favoured road building over water infrastructure, as men travelled more. With women on the councils, the priority skewed more to water.[17]

In other words, where water managers actively engage the actual users, or where the actual users *are* the water managers, water efficiency improves. As an ancillary benefit, where communities are obliged to work together to provide a common benefit, acrimony and tensions are reduced.

Soft-Path Strategy 5: Polluter Pays

This final strategy for water security we have already discussed in Chapter 3, on rivers. It is simple to state, if complex and politically fraught to implement: polluter pays. This applies to single households, businesses, and municipalities alike: if you put something into the water that degrades its quality, you pay the full cost of restoration. Singapore can do it. The EU is going that way. Why not the rest of the world?

The key to water efficiency, then, is a combination of clever regulation, price incentives, technological changes, decentralization, and citizen participation. Few of these are difficult to conceive or implement. Not all of them are expensive, though a few are. All of them have already been implemented somewhere, on some scale. All of them would help, even absent the others. In combination, they would buy us time.

Conclusion
Crossover Solutions

How to put all this together? How to put the soft path in the service of the hard path, and the hard path in the service of conservation? What is needed are crossover solutions. Engineer, meet ecologist. You're not that different, after all. You're both needed. You both live in this troubled place.

First, stop thinking of this as a single crisis. We'll never solve "it," but we may be able to solve "them"

I discussed this notion of reframing the debate in my brief Chapter 13. The key point is that conceiving of the planet's water woes as, well, the *planet's* water woes, gets you no closer to a solution.

A billion and a half people are without safe water, but you can't help them with a single policy prescription, no matter how well-intentioned and well-funded. The number is too large, too numbing — there is no possible program that can encompass the scale.

Water woes are local, and must be solved locally.

This doesn't mean that they can only be solved by local people, however. Outsiders (including the funding agencies, NGOs, *anyone*) can help. But — does it need to be said? — outsiders need to work *with* the locals, at their direction and to their needs. Nothing will get solved if outside agents go galloping off in all directions, at risk of contradicting and confusing each other, and at risk of alienating the people they ostensibly want to help.

It has long been a cliché of the environmental movement that we should "think globally but act locally." In this case, however, we should *think* locally too.

So narrow the focus. Not a billion but a number with more manageable boundaries. Think instead of the citizens of northern Mali who are living with water stress, think of the even smaller number in Yemen, think of the slum dwellers of Lagos, of the people who must perforce drink polluted water from the Mekong, of the farmers in southern India and the southwestern United States and North China and southeast Brazil who must share depleting aquifers and learn how to do more with less...On that scale, many of the problems become solvable, or at least more amenable to solution. Not easy to solve, but easier.

Second, think big and small at the same time

The rich-world water charities are widely derided by the technocrats of the development industries as being, at best, too little too late. In the technocratic vocabulary, "tokenism" has become the insult du jour. Do-gooders, charity porn, "poorism," Lord and Lady Bountiful. The insults are thick on the ground. But is this fair? A popular target is a charity called, plainly enough, charitywater.org, founded less than a decade ago by a refugee from the New York nightclub life, Scott Harrison. He has become a target because of his never-failing public cheerfulness and his unabashed use of the word "charity," which he defines as doing good but which in more cynical circles has come to mean something like preening self-esteem. In any case, charitywater. org has brought clean (or cleaner) water and sanitation to 4.6 million people in twenty-two countries, through 13,644 separate projects (hand-dug wells, drilled wells, rainwater catchments, gravity-fed systems, piped systems, water purification systems, bio-sand filters, spring projects, latrines). Sure, if you do the math, each project benefits on average a few more than 300 people, a very long way indeed from the billion-plus who need help. But you think those people aren't better off? Think villages, then. Think families. Think of Manya, the Gabbra woman from Kenya, and then think of her daughters, and small no longer seems pathetic.

The small is also enabling. A new well in a village that had been without one enables the villagers to use the new water to grow gardens near their homes, and secure their food supply — and possibly sell produce to earn a little money. Cumulatively, that reduces pressure on meagre government services. And it liberates a few people to learn things they would never have been free to learn otherwise.

But big is needed too. The point is, there is no reason why they can't coexist.

Here is one example from my own travels. The UN and the Aga Khan Foundation are building (or were — the jihadists put a stop to it) a sewage treatment plant for the city of Timbuktu, where before open-pit latrines were the norm. At the same time, a number of NGOs were helping individual families install simple bleach-based water purification systems, one household at a time. No one saw any conflict between the large and the small, between the notion of finally extending a piped-water grid to the masses of Kinshasa (something only a multinational can do) and showing individual families how to clean what water they now have (which small charities can do). They work together. One scale, the small one multiplied up, reduces the need for the other but doesn't do away with the need altogether.

Pakistan needs to build more large dams for water security, to be able to store more water for anxious times. The Pakistanis also need to conserve water on a household level — requiring education, instruction, propaganda even, and also penalties. Both approaches are necessary.

Third, make peace with your enemies

I was struck by how, in the Alberta oil patch, the word "environmentalist" has become a synonym for something deeply anti-social. Not just anti-business, but worse than that — someone who would be prepared to strip society's gears in the service of some mystical but ill-considered world view. And, of course, the antithesis is also true.

There are activists who profoundly believe that the corporate world will carelessly wreck the planet for competitive advantage.

These are the extremes. In the middle are organizations like the Natural Resources Defense Council and the Pacific Institute that are environmental stewards but also sympathetic to the profit motive and to business generally; and people like those of COSIA, on the business side, who don't want to pollute the world any more than the Council of Canadians does. The world's water problems will approach solution only if the two sides can connect. Conservation and engineering are two sides of the same coin. NGOs and private corporations should not be enemies — they have much in common, as the sensible ones would learn if they talked to each other. Social justice and economic development are not antithetical. The hard path and the soft path can work only if they are both implemented in consultation and not animosity. We need to shift vast quantities of water and also to conserve what we have. Both points of view are essential. So is understanding of "the other." Both sides should reject the strident extremes. Exclude the extremists by not listening to them. Closed minds should be encouraged to open, or risk being shunned. Listen only to the reasonable.

Fourth, co-opt the engineers and the hard scientists

Many of the soft-path solutions outlined in the previous chapter require high-end technology, entrepreneurial skills, and complex manufacturing, along with more basic research.

Fifth, enlist the environmentalists and their scientists

They are not hard to find. Their work shows up in the professional journals. The academies are full of people with their heads down, working to understanding the natural world, and with small tolerance for polemics. Ecology is a hard science too.

Sixth, abandon uncompromising, hardline positions on the bigger water issues

Dams, for instance. Bulk water transfers. Genetic modification. Privatization. None of these are necessarily good, but none of them are evil either. Far too much energy is being spent combatting things that need understanding instead.

By all means, demolish dams that no longer serve any purpose. Encourage run-of-stream hydro developments instead of static reservoirs. Make sure that dams are built only with proper consultation with and compensation for those most affected. Properly designed and built, massive dams give water security where there was none before and generate large amounts of power that is less destructive than any of the alternatives. Yes, it is important to protect vulnerable and poor peasant societies. But it is not acceptable for rich countries and rich NGOs to pursue policies whose effect would be to preclude development and keep them poor peasants forever.

By all means, keep water within its natural basin wherever possible. And don't allow bulk water transfers when the consequences are overwhelmingly negative. But, as discussed in Chapter 8, there are hundreds of perfectly acceptable pipelines or canals, in every continent, that have yet to be built. Don't assume that every such transfer is destructive. Just work to see that it isn't.

By all means, forbid the patenting of seed varieties that are proprietary and designed to work with, or against, certain pesticides and herbicides. Don't allow companies to lock farmers into buying seeds from a single company and then to forbid them to replicate those seeds. Don't allow patents on any life forms, no matter how derived. As I said in Chapter 9, on farming, we should, however, now accept that genetic manipulation can be beneficial and should approve and encourage attempts to create transgenic varieties that increase yields, confer resistance to viral diseases, and thrive in poor farming conditions — and from which farmers can derive their own seeds. When discussing GM, then, one side needs to scale down its arrogance, the other its paranoia. Both sides should get a grip.

By all means, watch that water privatization doesn't exploit the vulnerable. This is easiest of all these injunctions. Get over the suspicion, and design better deals.

Seventh, enlist the private sector

Recognize the virtues of private corporations while watching for their sins — recognize their skills and also their limitations. They shouldn't be allowed to police themselves, but if they are properly regulated, private corporations are nimble and technologically adept, and have access to vast amounts of capital from markets across the globe. The opposition needs to understand that the millions of poor people who have access only to limited, unclean water supplies can be served only by privatization — nobody else has the capability, or the money, or the will. In many poor parts of the world, the real question is not whether private interests should be allowed to participate in providing water or whether private interests will exploit the poor, it is whether the poor get any water at all. The proper approach is not to be against private involvement. The proper approach is to see that it works as well as it can, for all parties.

Eighth, take a harder line on polluters

Fixing pollution is, or at least should be, easy. We know how to do it. The technology is to hand. The policy prescriptions are to hand. There are examples of successes to hand too — see the Rhine and the Hudson, and to a lesser degree the Danube, and the work of the EPA in the United States and the EU's water directive initiative. It will be expensive, but then pollution is expensive too, necessitating vast arrays of palliative technologies to make water safe. We don't even need to spend a lot of time cleaning up historic pollution because nature will do a lot of that for us. What we need to do is stop polluting further — and to punish people who do. Do a proper cost-accounting of the burdens of pollution and make the people who dirty the water clean it up.

It really is that simple. So why don't we just do it?

In the rich world, at least in part, we do. To a surprising degree, though, not nearly enough. Of course, that's where it stops being simple.

In the poor world, matters are dire. In the poor world, the corporate sector isn't usually the villain, though there are well-known exceptions. The villains are government ineptitude and carelessness, more urgent priorities for scarce resources, massive poverty, lack of education, and rapidly increasing populations. The burdens of poverty and the wretched quality of available services mean that governments simply don't have the money to fix even the obvious problems, never mind the hidden ones. Money for remediation is just not available.

In the rich world, villainy has a wider spread. In the rich world, a corporation that makes, say, car batteries, could until recently pour toxic effluent into, say, the Love Canal (to take a well-known recent-history example). The company did it because it could and because the costs of polluting have always been appallingly low. This always baffled activists — *Don't the executives have children? Don't they care? How can they possibly not care?* But these are the wrong questions. The executives permit pollution, or at least did in the past, because, in general, it went away without any effort on their part and it cost them nothing. It ran away into the rivers and thence to the sea, and then it was not their problem anymore. They had children, sure, but *their* children were not dying of polluted water. Municipal filtration and purification systems keep their children's water safe, and the costs of these systems are not built into corporate life but are seen as externalities. They become society's costs. Further, corporations permit pollution because there is really no external regulatory authority to tell them not to, though this is obviously no longer as true as it once was. They polluted because it didn't cost them to do so. Until recently, there was no profit in conservation.

That's why I say fixing pollution is easy — in the rich world. The law we can write can be simple in its essence: the water going out of your facility should be as clean as the water coming in. If it isn't, you pay the full cost of remediation. (Of course, definitions of "full cost" will keep tribes of lawyers busy for decades, but it is still worth attempting.) The principle is easy to see: if there are more than a hundred chemicals in the Great Lakes and we don't even know what seventy of them are, it doesn't seem unfair to insist that those who put them there either get them out or refrain from putting any more in. The Germans and many other European countries have already moved in this direction, with significant results.

Citizen vigilance and political activism are having their effects. There remain grounds for optimism. Corporate buy-in is still small but it is growing — all those relentless activist questions are resonating in boardrooms whose executives, who are really not the conscienceless monsters of ecological legend, really do have children and in many cases really do care, or can be led to care. In the rich world too, organizations like the Natural Resources Defense Council, Riverkeepers, and more controversially, Greenpeace are beginning to count their victories.

What can be done about it elsewhere? The gloomy answer is that in one real sense, not much. Poor-world pollution is not a water issue alone; it is a product of crushing poverty, inept national institutions, massive population increases, and a knowledge base in which anti-pollution technologies were adequate for sparse rural populations but catastrophic for crowded urban slums. Only increasing affluence can change this lamentable state of affairs, and, as is evident from the case of China, newly achieved affluence actually makes things worse, at least at first.

The less gloomy answer is to focus on nibbles rather than bites. Pollution issues are tractable to small initiatives. Every little bit really does help. In this sense, foundations and aid groups can make a significant difference to a fairly large number of individual lives and villages.

What about money? Can the rich world help? Western NGOs suggest, with some plausibility, that it is the rich world's moral duty to help the poor one. Not only because the rich world causes a good deal of pollution but also because millions dying of entirely preventable diseases is a global disgrace. And also, though this is generally unstated, because pollution damages human health (risking border-crossing pandemics) and exacerbates global instability (leading to environmental refugees). But how do you translate a moral duty into a practical program? Block grants and aid allocations to inept governments won't help; the money will mostly just disappear.

There is no one "real" answer, then. But in a strategic sense, the best answer is to create the conditions where the citizens affected bring about the political conditions where pollution is no longer acceptable. It means massive investments in education. It means schools, schools, schools. It means drumming out the kleptocrats. It means supporting the kind of development proposals that would be most likely to help poor countries become richer. Like massive infrastructure projects, energy, and transportation. Highways. Hydroelectric schemes. Dams, even. Give them some economic freedom and they'll fix it themselves. Eventually. Even China. It hasn't gone unnoticed by environmentalist groups worldwide, and inside China itself, that by 2014 the country was, finally, getting serious about pollution. On the last day of 2014, the Xinhua news agency reported that six companies had been fined $26 million for polluting rivers on which their factories depended, a record penalty for the country. Even better, fourteen people were jailed for their malfeasance.

It is taking the already-rich world long enough, but it is, generally, getting there.

Ninth, spend where you have to

Remediation is going to cost money, very large sums of money, but we can see what happens with inaction — the American Southwest reaches critical stages, so do parts of Brazil, so do many other parts of the world, and infrastructure is crumbling pretty much everywhere. We need to spend money to fix leaks, to find new sources, to protect those sources we already have.

We need to spend money to invent new ways of managing water.

And then, we need new technologies: bioremediation, wastewater treatment, better desalination, new appliances, smarter meters, and pipe systems — this is already happening: encourage it.

Tenth, treat thrift as the highest virtue

Los Angeles is probably not the first community that comes to mind when contemplating the beguiling notion of thrift (or conservation, in water terms). Not when, in the fourth year of an ongoing drought, members of the 0.001 percent are tearing down mere 4,500-square-metre houses to build 7,500-square-metre ones... "So long, mega-mansion, say hello to the giga-mansion," in Peter Haldeman's phrase.[1]

But this wretched-excess ostentation is not the whole story. It isn't even the main story. The real story is that Los Angeles now uses less water than it did in 1970, not just per capita but in total, though the population has grown by more than a third. The real story is that the city has won a water sustainability award given by the US Water Alliance. The real story is that the city has built a new water-treatment wetland where a bus yard used to be. The real story is that the metropolitan area, prodded by the prolonged drought and facing restrictions on importing water from elsewhere, has created a slew of new projects that "will treat polluted and even sewage water, capture rainwater, store water in aquifers, and use (or reuse) all of it, often while mimicking or supporting natural processes. The area's water administrators who, until recently, thought of watersheds as merely rural concerns now recognize that even in Los Angeles, all living things are linked by their common water course and that its proper management is essential to the administrators' success."[2] This from Jacques Leslie in the *New York Times*. Late in 2014, the city issued a directive that imported water be cut in half within ten years. The study I quoted in the previous chapter, by the Natural Resources Defense Council and the Pacific Institute, found that a fully developed stormwater capture system in Los Angeles could add 381 million cubic metres to the city's supply — a good part of the total consumption that is currently 724 million cubic metres. This stormwater capture system is already half built. It will make the notion of importing water from long distances seem like a profligate waste of energy and resources.

Now to persuade the mega-rich that one swimming pool will do...

Eleventh, cling to the notion that water problems are a management issue, not a resource issue

Because that really is the good news about water. There is water enough. We have skills ample enough to provide it to everyone alive and soon to be born; we have the technological expertise to clean water to a fare-thee-well; we have the energy to move it wherever we want in whatever quantity we want, preferably, of course, within river basins. Water is not the issue. Our dysfunctional political systems are. Those should be easy to fix. They're not — but they should be.

Twelfth, design a basic municipal water system that can serve as a template in many places

Ample examples have been given in the foregoing chapters. Enough water left alone for the environment. Subsidized (cheap) water for the poor — "the first glass is free." Meter everything. Mandate efficiency. Price for conservation — make water *much* more expensive. Escalate prices sharply when consumption increases. End industry's free ride, and farming's too. Penalize waste, and penalize polluters. Recycle everything.

Thirteenth, don't think you have failed if you fail to solve everything

You won't solve everything. Large-scale social engineering is too complicated to think otherwise. But the water world can certainly get better.

Notes

Chapter 1

1. Report from Reuters, November 29, 2014.

2. Adriana Brasileiro, "São Paulo Running Out of Water as Rain-Making Amazon Vanishes," *Reuters*, October 24, 2014.

3. Quoted by Thomas Friedman, *New York Times*, November 4, 2014.

4. These quotes are from Peter H. Gleick and Meena Palaniappan, "Peak Water Limits to Freshwater Withdrawal and Use," *Proceedings of the National Academy of Sciences* 107, no. 25 (June 22, 2010): 11155–62.

5. Taikan Oki and Shinjiro Kanae, "Global Hydrological Cycles and World Water Resources," *Science* 313, no. 1068 (2006), doi: 10.1126/science.1128845.

6. From a 2003 IRN paper, "A Crisis of Mismanagement: Real Solutions to the World's Water Problems."

7. These numbers from Oki and Kanae, "Global Hydrological Cycles and World Water Resources."

Chapter 2

1. For more detail, see "Groundwater Use for Irrigation — A Global Inventory," by a shopping list of authors from three institutions, the Institute for Crop Science at the University of Bonn, the FAO in Rome, and the Institute of Physical Geography, University of Frankfurt/Main: S. Siebert, J. Burke, J.M. Faures, K. Frenken, J. Hoogeveen, P. Döll, and F.T. Portmann, *Hydrology and Earth System Sciences Discussions* 7, 3977–4021 (2010), doi: 10.5194/hessd-7-3977-2010.

2. nature.com/news/reservoir-deep-under-ontario-holds-billion-year-old-water-1.12995.

3. The world's largest aquifers, in no hierarchy of size, are these: 1. Nubian aquifer system (NAS); 2. Northwest Sahara aquifer system (NWSAS); 3. Murzuk-Djado basin; 4. Taoudeni-Tanezrouft basin; 5. Senegalo-Mauritanian basin; 6. Iullemeden-Irhazer aquifer system; 7. Chad basin; 8. Sudd basin (Umm Ruwaba aquifer); 9. Ogaden-Juba basin; 10. Congo Intracratonic basin; 11. Northern Kalahari basin; 12. Southeast Kalahari basin; 13. Karoo basin; 14. Northern Great Plains/Interior Plains aquifer; 15. Cambro-Ordovician aquifer system; 16. California Central Valley aquifer system; 17. High Plains-Ogallala aquifer; 18. Gulf Coastal Plains aquifer system; 19. Amazonas basin; 20. Maranhao basin; 21. Guarani aquifer system; 22. Arabian aquifer system; 23. Indus basin; 24. Ganges-Brahmaputra basin; 25. West Siberian artesian basin; 26. Tunguss basin; 27. Angara-Lena artesian basin; 28. Yakut basin; 29. North China Plain aquifer system; 30. Songliao basin; 31. Tarim basin; 32. Parisian basin; 33. East European aquifer system; 34. North Caucasus basin; 35. Pechora basin; 36. Great artesian basin; 37. Canning basin.

4. See http://www.nrcan.gc.ca/evaluation/reports/2013/11142 for how to find this data.

5. Report by Matt McGrath, April 20, 2012, "'Huge' Water Resource Exists under Africa."

6. The report, "Quantitative Maps of Groundwater Resources in Africa," was published in the

journal *Environmental Research Letters* 7 no. 2 and is available on the British Geological Survey website, http://www.bgs.ac.uk/research/groundwater/international/africanGroundwater/maps.html; Bonsor was quoted by the BBC.

7. See Jim Yardley, "Beneath Booming Cities, China's Future Is Drying Up," *New York Times*, September 28, 2007.

8. From scientists from Fliders University, Adelaide: Vincent E.A. Post, Jacobus Groen, Henk Kooi, Mark Person, Shemin Ge, and W. Mike Edmunds, "Offshore Fresh Groundwater Reserves as a Global Phenomenon," *Nature* 504 (December 5, 2013), doi: 10.1038/nature12858.

9. Quoted by Neena Satija, "What's the Magic Number on Texas Water Needs?" *Texas Tribune*, May 8, 2014.

10. See the interesting piece, "Wells Dry, Fertile Plains Turn to Dust," by Michael Wines in *New York Times*, May 19, 2013.

11. Blog by Annabel Symington, "The Guarani Aquifer: A Little Known Water Resource in South America Gets a Voice," April 13, 2010, htttp://blogs.ei.columbia.edu/2010/04/13/.

12. Global Water Forum, *The Agreement on the Guarani Aquifer: Cooperation without Conflict*, September 2, 2013, http://www.globalwaterforum.org/2013/09/02/the-agreement-on-the-guarani-aquifer-cooperation-without-conflict.

Chapter 3

1. Africa: the Congo, Niger, and Nile, with a combined annual runoff of 1,982 cubic kilometres. South America: the Amazon, Paraná, Orinoco, and Magdalena, with a combined runoff of 8,829 cubic kilometres, much of it due to the mighty Amazon. Asia: the Ganges, Yangtze, Yenisei, Lena, Mekong, Irrawaddy, Ob, Chutsyan, Amur, Indus, and Salween/Nu, with a flow of 5,722 cubic kilometres. North America: the Mississippi, St. Lawrence, Mackenzie (Canada's longest at 4,241 kilometres), Columbia, and Yukon, with a combined flow of 1,843 cubic kilometres. Europe: the Danube and Volga, with a combined flow of 468 cubic kilometres.

2. News item in *Science*, October 17, 2014, reporting a study published in *Geophysical Research Letters.*

3. This from a report, "China's Water Shortage Could Shake World Food Security" by Lester Brown and Brian Halweil in *WorldWatch* magazine, as early as 1998.

4. Tao Tao and Kunlun Xin, "A Sustainable Plan for China's Drinking Water," *Nature* 511 (31 July 2014): 527–528; doi:10.1038/511527a.

5. Stephen Luby, "Water Quality in South Asia" *Journal of Health, Population and Nutrition*, June 2008, 26, no. 2: 123–24. Luby is, among other academic appointments, head of the Program on Infectious Diseases and Vaccine Sciences at the International Centre for Diarrhoeal Disease Research, Bangladesh.

6. Wijarn Simachaya, "MRC-Water Quality Monitoring Network (WQMN)," October 2013.

7. Canadian Press report by Dene Moore, August 2014.

8. Quoted in Gordon Hoekstra, "Residents Affected by Mount Polley Dam Failure Hope Panel Report Prompts Charges," *Vancouver Sun*, November 15, 2014.

9. *Halifax Chronicle Herald*, August 6, 2014.

10. Dirk Meissner, "Victoria Sewer Dispute Hits the Fan as Washington State Urges B.C. Intervene," Canadian Press, June 11, 2014.

11. Kurt Hollander, "Mexico City: Water Torture on a Grand and Ludicrous Scale," *Guardian*,

February 5, 2014.

12. Report from the Commission to the Council and the European Parliament, Brussels, 4.10.2013 COM (2013) 683 final.

13. ILEC/Lake Biwa Research Institute, eds., *1988–1993 Survey of the State of the World's Lakes*, vols. 1–4 (Nairobi: International Lake Environment Committee, Otsu, and United Nations Environment Programme).

14. David W. Schindler, R.E. Hecky, D.L. Findlay, M.P. Stainton, B.R. Parker, M.J. Paterson, K.G. Beaty, M. Lyng, and S.E.M. Kasian, "Eutrophication of Lakes Cannot Be Controlled by Reducing Nitrogen Input: Results of a 37-Year Whole-Ecosystem Experiment," *Proceedings of the National Academy of Sciences of the United States of America* 105, 32 (2008): 11254–58; doi: 10.1073/pnas.0805108105.

15. This from Michael Wines's excellent survey of the lake, "Behind Toledo's Water Crisis, a Long-Troubled Lake Erie," *New York Times*, August 4, 2014.

16. Two good works on eutrophication are David Schindler and John R. Vallentyne, *Overfertilization of the World's Freshwaters and Estuaries* (Edmonton: University of Alberta Press, 2008), and M. Nasir Khan and F. Mohammad, "Eutrophication: Challenges and Solutions," *Eutrophication: Causes, Consequences and Control*, ed. A.A. Ansari (Dordrecht: S.S. Gill Springer Science+Business Media, 2014); doi:10.1007/978-94-007-7814-6_5.

17. Lindsay Crouse, "River Grime? Triathletes Are Swimming in It," *New York Times*, August 3, 2014.

18. *Environmental Science & Technology*, September 2014. See also Pesticide National Synthesis Project, http://water.usgs.gov/nawqa/pnsp.

19. Quoted by Willemien Groot on Radio Nederland, August 21, 2006, "Cleaning Up the Filthy River Rhine."

20. I.P. Zaretskaya, "Water Availability and Use in the Danube Basin," and "State of the Art: Expected Water Availability and Water Use in the Danube Basin," paper delivered to a UNESCO water conference, Paris, 2006.

21. Some of this through personal correspondence with Philip Weller.

22. *Science* 347, no. 6219 (16 January 2015), doi:10.1126/science.1255641, and Carl Zimmer, "Ocean Life Faces Mass Extinction, Broad Study Says," *New York Times*, January 15, 2015.

23. C. Nellemann and E. Corcoran, eds., *Our Precious Coasts — Marine Pollution, Climate Change and the Resilience of Coastal Ecosystems*, United Nations Environment Programme, 2006.

24. http://www.riverfoundation.org.au/.

25. ILEC, *Managing Lakes and Their Basins for Sustainable Use: A Report for Lake Basin Managers and Stakeholders*, International Lake Environment Committee Foundation, 2005, http://www.worldlakes.org/uploads/LBMI_Main_Report.pdf.

Chapter 4

1. Catherine Brölmann, "International Law as Tool for Global Water Governance," http://www.thebrokeronline.eu/Authors/Broelmann-Catherine.

2. Some of this was recounted in my earlier book, *Water: The Fate of Our Most Precious Resource* (Toronto: McClelland and Stewart, 2003).

3. Joanna L. Robinson, *Contested Water: The Struggle against Water Privatization in the*

United States and Canada (Cambridge, MA: MIT Press, 2013).

4. Maggie Black and Jannet King, *The Atlas of Water: Mapping the World's Critical Resource*, (Berkeley: University of California Press, 2009), 92.

5. Terry Anderson and Pamela Snyder, *Water Markets* (Washington, DC: Cato Institute, 1997), 54.

6. Peter Gleick, "Whose Water Is It? Water Rights in the Age of Scarcity," SFGate, August 2, 2009. http://blog.sfgate.com/gleick/2009/08/02/whose-water-is-it-water-rights-in-the-age-of-scarcity/. Reprinted by permission of the author.

7. Neena Satija, "Texas Groundwater Districts Face Bevy of Challenges," *Texas Tribune*, August 29, 2013.

8. There's an interesting discussion of this issue in Anderson and Snyder, *Water Markets*, 185.

9. Ramona Giwargis, "Merced Supervisors to Consider First Draft of Groundwater Ordinance," *Modesto Bee*, October 19, 2014.

10. Felicity Barringer, "Desperately Dry California Tries to Curb Private Drilling for Water," *New York Times*, August 31, 2014.

Chapter 5

1. *Blue Covenant*, repeated (slightly reworded) in a piece for *American Prospect*, June 2008, https://prospect.org/article/where-has-all-water-gone.

2. Fredrik Segerfeldt, *Water for Sale: How Business and the Market Can Resolve the World's Water Crisis* (Washington, DC: Cato Institute, 2005), 106, 6.

3. Paul Farrell, "Water Is the New Gold, a Big Commodity Bet," *MarketWatch*, July 24, 2012.

4. Emily Achtenberg, *Rebel Currents*, June 6, 2013, nacla.org/column/7334, originally published in ReVista, *Harvard Review of Latin America* 12, no. 2 (Winter 2013).

5. Shultz was writing in *Yes! Magazine*. His piece was called "The Cochabamba Water Revolt, Ten Years Later," April 10, 2010. He was also co-editor of *Dignity and Defiance, Stories from Bolivia's Challenge to Globalization* (University of California Press), and was in Cochabamba throughout the turmoil.

6. Broadcast January 26, 2009. His commentary was later repeated in the *Guardian* and on the Latin American news website Upside Down World.

7. Quoted in Anderson and Snyder, *Water Markets*, 50.

8. In conversation with the author, 2014.

9. Quoted in McKenzie Funk, *Windfall: The Booming Business of Global Warming* (New York: Penguin, 2014), 118.

10. Paul B. Farrell, "Water is the New Gold, a Big Commodity Bet."

11. "Ebb and Flow: Competition Is Being Drip-Fed into the Water Industry," *Economist*, November 22, 2014.

12. Dickerson's story is well told in McKenzie Funk's entertaining book *Windfall*.

13. Peter H. Gleick et al., *The World's Water*, vol. 7 (Washington, DC: Island Press, 2012), 25.

14. This last example from Charles Fishman's *The Big Thirst: The Secret Life and Turbulent Future of Water.* (New York: Simon and Schuster, 2011), 267.

15. A good survey of this issue is by Neena Satija of the *Texas Tribute* arm of the *New York Times*, "Aquifer Is No Quick Fix for Central Texas Thirst," September 11, 2014.

16. Mike Esterl, "U.S. Water Privatization Fails to Pan Out," *Wall Street Journal*, June 26,

. 2006.

17. "Value Diluted: Water Is a Growing Business Problem: Many Companies Haven't Noticed," *Economist*, November 8, 2014.

18. Martin Pigeon, "From Fiasco to DAWASCO: Remunicipalisation Problems in Dar es Salaam, Tanzania," in *Remunicipalisation: Putting Water Back into Public Hands*, ed. Martin Pigeon, David A. McDonald, Olivier Hoedeman, and Satoko Kishimoto (Amsterdam: Transnational Institute, 2012). Full text available online at http://www.municipalservicesproject.org/publication/remunicipalisation-putting-water-back-public-hands.

19. Funk, *Windfall*, 132.

20. David Lewis Feldman, "Australia's Drought: Lessons for California," *Science* 343, no. 6178 (March 28, 2014): 1430.

21. Gleick, *The World's Water*, vol. 7, 96.

22. Much of this, as well as some of the other examples cited, is from the Remunicipalization Tracker.

23. Victoria Collier, "Deep Questions Arise over Portland's Corporate Water Takeover," Truthout website, January 7, 2015.

24. http://www.municipalservicesproject.org/.

25. "Puerto Rico's Debt-Ridden Water Authority Cuts off Non-Payers," *Latin American Herald Tribune*, www.laht.com/article.asp?ArticleId=774453&CategoryId=14092.

26. John Eligon, "Detroit Threatens to Cut Water Service to Delinquent Customers," *New York Times*, March 25, 2014.

27. Gleick, *The World's Water*, vol. 7, 36.

28. Briscoe said this to me but had earlier made the same point in his *Water Policy* interview. See Chapter Six, note 2.

29. Many of these very sensible conclusions are from Andrew Nickson and Claudia Vargas, "The Limitations of Water Regulation: The Failure of the Cochabamba Concession in Bolivia," *Bulletin of Latin American Research* 21, no. 1 (2002): 99–120.

30. Joseph Berger, "Desalination Plan Draws Ire in Rockland County," *New York Times*, November 13, 2014.

31. Ellen Dannin, "If It Sounds Too Good . . . What You Need to Know, but Don't, about Privatizing Infrastructure," Truthout website, October 31, 2013.

32. Marq de Villiers, *Our Way Out: First Principles for a Post-Apocalyptic World* (Toronto: McClelland and Stewart, 2001), 274.

Chapter 6

1. John McPhee, "Farewell to the Nineteenth Century," *New Yorker*, September 27, 1999.

2. The Briscoe material in this chapter comes partly from "Overreach and Response: The Politics of the WCD and Its Aftermath," *Water Alternatives* 3, no. 2: 399–415; partly from an interview he gave to the editor of *Water Policy*, Jerome Delli Priscoli, published in *Water Policy* 13 (2011): 146–60; and partly from a conversation with the author from his Harvard office (July 2014).

3. The McCully quotations in the following paragraphs are from Patrick McCully, "The Use of a Trilateral Network: An Activist's Perspective on the Formation of the World Commission on Dams," *American University International Law Review* 16, no. 6 (2001): article 3.

4. Quoted by Briscoe, "Overreach and Response: The Politics of the ICD and Its Aftermath," *Water Alternatives* 3, no. 2 (2010): 399.

5. Letter to *Foreign Policy*, September 2004.

6. *Times of India*, June 13, 2014, timesofindia.indiatimes.com/.

7. Ibid.

8. "Large Dams Just Aren't Worth the Cost," *New York Times*, August 24, 2014.

9. A. Ansar et al., "Should We Build More Large Dams? The Actual Costs of Hydropower Megaproject Development," *Energy Policy* (2014), http://dx.doi.org/10.1016/j.enpol.2013.10.069i.

10. Sebastian Mallaby, *The World's Banker* (New York: Penguin, 2006), 7.

11. World Energy Council, http://www.worldenergy.org/data/resources/resource/hydropower/.

12. This from a good piece by Andrew Revkin, "Can Bhutan Achieve Hydropowered Happiness?" *New York Times,* December 10, 2013.

13. Amelia Urry, Lisa Hymas, and Sara Bernard, "'Night Moves' Is Why People Hate Environmentalists," Grist website, September 5, 2014.

14. Rachel Smolker and Almuth Ernsting, "Abundant Clean Renewables? Think Again!" Truthout website, November 16, 2014.

15. "World Bank Acknowledges Shortcomings in Resettlement Projects, Announces Action Plan to Fix Problems," World Bank press release, March 4, 2015.

16. This paragraph is from Marq de Villiers, *Dangerous World* (New York: Penguin, 2008).

17. Ivan B.T. Lima, Fernando M. Ramos, Luis A.W. Bambace, Reinaldo R. Rosa, "Methane Emissions from Large Dams as Renewable Energy Resources: A Developing Nation Perspective," *Mitigation and Adaptation Strategies for Global Change* 13, no. 2 (February 2008).

18. Quoted in Keith Schneider's "Uttarakhand's Furious Himalayan Flood Could Bury India's Hydropower Program," April 2, 2014. Schneider is senior editor for the Circle of Blue website.

19. Andrew Jacobs, "Plans to Harness Chinese River's Power Threaten a Region," *New York Times*, May 4, 2013.

20. Marq de Villiers and Sheila Hirtle, *Into Africa: A Journey through the Ancient Empires* (Toronto: Key Porter Books, 1999).

21. International Rivers press release, June 19, 2012.

Chapter 7

1. Cherry's quotes, as well as Russell Gold's, are from a fracking conference held in Toronto, May 29, 2014, at the Munk School of Global Affairs, part of its Program on Water Issues.

2. Tillerson's notorious quote has been widely reported by the *New Yorker*, the *New York Times*, *American Scientist*, and many other publications and websites. See, for example, Ian Urbana's piece in the *New York Times*, August 3, 2011, titled "A Tainted Water Well, and Concern There May Be More."

3. *Globe and Mail*, May 1, 2014.

4. Louis Sahagun, "U.S. Officials Cut Estimate of Recoverable Monterey Shale Oil by 96%," *Los Angeles Times*, May 21, 2014.

5. http://articles.philly.com/2013-05-14/news/39231025_1_marcellus-shale-terry-engelder-natural-gas.

6. Asjylyn Loder and Isaar Arnsdorf, "Majority of U.S. Companies Inflating Shale Reserves," *Bloomberg News*, October 10, 2014.

7. These numbers from CERES, a consulting firm that pushes for sustainable business practices.

8. For more on this, see Jim Malewitz and Neena Satija, "In Oil and Gas Country, Water Recycling Can Be an Extremely Hard Sell," *New York Times*, November 21, 2013.

9. "Clean That Up: Environmental Technology; A Combination of Two Desalination Techniques Provides a New Way to Purify the Water Used in Fracking," *Economist*, November 30, 2013.

10. "Study of the Potential Impacts of Hydraulic Fracturing on Drinking Water Resources: Progress Report," http://www2.epa.gov/hfstudy.

11. Roger Drouin, "Fracking Unfocus: How the EPA's Long-Awaited Hydraulic Fracturing Study Could Miss the Mark," Truthout website, November 18, 2013.

12. Bill McKibben, "Bad News for Obama: Fracking May Be Worse Than Burning Coal," *Mother Jones*, September 8, 2014, http://www.motherjones.com/environment/2014/09/methane-fracking-obama-climate-change-bill-mckibben.

13. Dana R. Caulton et al., "Toward a Better Understanding and Quantification of Methane Emissions from Shale Gas Development," *Proceedings of the National Academy of Sciences* 111, no. 17: 6237–42, doi: 10.1073/pnas.1316546111.

14. "Oklahoma Earthquakes Induced by Wastewater Injection by Disposal Wells, Study Finds," *Science* 345, no. 6192 (July 4, 2014): 13, 14.

15. Susan Hough, quoted in a news report by Alexandra Witze, "Man-Made Quakes Shake the Ground Less than Natural Ones: Seismic Danger from Oil and Gas Operations May Be Overestimated," *Nature*, August 2014, doi:10.1038/nature.2014.15742. Hough's original study was published in *Journal of the Seismological Society of America*.

16. Erin Kelly et al., "Oil Sands Development Contributes Elements Toxic at Low Concentrations to the Athabasca River and Its Tributaries," *Proceedings of the National Academy of Sciences* 107, no. 37 (July 2, 2010): 16178–83, doi: 10.1073/pnas.1008754107.

17. Keith Schneider and Sam Kean, "Tar Sands Oil Production, an Industrial Bonanza, Poses Major Water Use Challenges," Circle of Blue website, August 10, 2010.

18. The thirteen member companies of COSIA are Canadian Natural Resources, Nexen, Syncrude, BP, Cenovus, Teck, Devon, Statoil, Imperial, Total, Shell, ConocoPhilips, and Suncor.

Chapter 8

1. Steven Erie, quoted in "A Hundred Years of Soggy Tubes: California's Largest City Salutes the Source of Its Growth," *Economist*, November 9, 2013.

2. Peter H. Gleick et al., *The World's Water*, vol. 8 (Washington, DC: Island Press, 2014): 123ff.

3. Some of this from a survey in the *Economist*, October 12, 2013, titled "Rivers Are Disappearing in China; Building Canals Is Not the Solution."

4. Zhijun Ma et al., "Rethinking China's New Great Wall," *Science*, November 21, 2014.

5. Indian project via an editorial in *Science*, July 11, 2014.

6. Gonce material in correspondence with the author.

7. Rocky Casale and Reyhan Harmanci, "The Cold Rush," *Modern Farmer*, January 7, 2014.

8. Cran material in correspondence with the author.

9. This quote from Mulroy is from Fishman, *The Big Thirst*, 85.

10. Quoted by Eliza Barclay, "Alaska Town Eyes Shipping Water Abroad," *National Geographic News*, June 25, 2010.

Chapter 9

1. David Molden, ed., *Water for Food, Water for Life: A Comprehensive Assessment of Water Management in Agriculture* (London: Earthscan and Colombo, Sri Lanka: International Water Management, 2007).

2. In Sandra Postal, *Water: Adapting to a New Normal* (Santa Rosa, CA: Post Carbon Institute, 2010), a Post Carbon Reader Series publication.

3. http://www.ars.usda.gov/Aboutus/docs.htm?docid=10201.

4. Lester Brown, *Plan B: Rescuing a Planet Under Stress and a Civilization in Trouble* (Washington, DC: Earth Policy Institute, 2003). Reprinted by permission of the author and publisher.

5. Earth Policy Institute, *Raising Water Productivity to Increase Food Security*, June 22, 2012.

6. "Beyond Drip Irrigation to Water Fields in Dry Land Areas: An Interview with David Bainbridge," by Janeen Madan, research intern with the Nourishing the Planet project. The interview is on the website WorldEnvironment.tv.

7. Mark Bittman, "A Sustainable Solution for the Corn Belt," *New York Times*, November 18, 2014.

8. Research published by the Geological Society of America. See David R. Montgomery, "Is Agriculture Eroding Civilization's Foundation?" *GSA Today* 17, no. 10, doi: 10.1130/GSAT01710A.

9. D. Renault and W.W. Wallender, "Nutritional Water Productivity and Diets," *Agricultural Water Management* 45, no. 3 (2000).

10. For more on virtual water, see Daniel Renault's paper for a UNESCO workshop in 2002 titled "Value of Virtual Water in Food: Principles and Virtues."

11. The Post Carbon Institute published the *Post Carbon Reader* in 2010. Sandra Postel's chapter was called "Preparing for a Water-Limited World."

12. See http://www.csiro.au.

13. Tracy McVeigh, "Humble Spud Poised to Launch a World Food Revolution," *Guardian Weekly*, November 7, 2014.

14. de Villiers, *Our Way Out*, 210.

15. Joel K. Bourne, "The Global Food Crisis — The End of Plenty," *National Geographic*, June 2009, https://standeyo.com/NEWS/09_Food_Water/090526.end.of.plenty.html.

16. Krishna N. Das and Mayank Bhardwaj, "Modi Bets on GM Crops," Reuters, February 23, 2015.

17. Brown, *Plan B*. Reprinted by permission of the author and publisher.

Chapter 10

1. I have a clipping pinned to my office wall from one of these only-in-America contrarians, asserting that it was his God-given right to install however many and whatever showers he wanted in his house even if they used up as much water as a fire hydrant, and be damned to interfering bureaucrats who wanted to stop him. "It's my water and I'll use as much as I want to," he declared. The letter was sent to the editors of *Fine Homebuilding* magazine, which had recently been arguing for greener housing.

2. Justin Sheffield, Eric F. Wood, and Michael L. Roderick, "Little Change in Global Drought

over the Past 60 Years," *Nature* 491: 435–38, doi:10.1038/nature11575; Aiguo Dai, "Increasing Drought under Global Warming in Observations Ad Models," *Nature Climate Change* 3 (2013): 52–58, doi:10.1038/nclimate1633.

3. Justin Gillis "Science Linking Drought to Global Warming Remains Matter of Dispute," *New York Times*, February 16, 2014.

4. Andrew Revkin, "A Climate Analyst Clarifies the Science behind California's Water Woes," *New York Times*, March 6, 2014.

5. "Insurers' Disaster Files Suggest Climate Is Culprit," *Nature* 441 (June 8, 2006): 674–75, doi: 10.1038/441674a8.

6. Quotations in this paragraph and the next are from "Water and Climate Change: Understanding the Risks and Making Climate-Smart Investment Decisions," World Bank position paper #52911, November 2009.

7. Richard G. Taylor, Martin C. Todd, Lister Kongola, Louise Maurice, Emmanuel Nahozya, Hosea Sanga , and Alan M. MacDonald, "Evidence of the Dependence of Groundwater Resources on Extreme Rainfall in East Africa," *Nature Climate Change* 3 (2012): 374–78, doi: 10.1038/nclimate1731.

8. News report in *Nature*, November 13, 2014.

9. Some of this is discussed in my earlier book, *Windswept*, published in 2006 by Walker and Company in the United States and by McClelland and Stewart in Canada.

10. "In Spain, Water Is a New Battleground," *New York Times*, June 3, 2008.

11. http://www.worldbank.org/en/news/video/2013/02/05/melting-glaciers-slow-disaster-andes.

12. Ibid.

13. "Highest-Elevation Glaciers Keep Their Cool: A Case Study from the Nepal Himalaya," Cires.colorado.edu/science/spheres/snow-ice/glaciers.html.

14. This, and certain other facts in this section, are from the comprehensive survey and assessment of glaciers "Glacier Retreat: Reviewing the Limits of Human Adaptation to Climate Change," by Ben Orlove in *Environment* magazine, May–June 2009.

15. Porter Fox, "The End of Snow?" *New York Times*, February 7, 2014.

16. See survey by Michael Wines, "Climate Change Threatened to Strip the Identity of Glacier National Park," *New York Times*, November 22, 2014.

17. March 28, 2009.

18. Carling C. Hay, Eric Morrow, Robert E. Kopp, and Jerry X. Mitrovica, "Probabilistic Reanalysis of Twentieth-Century Sea-Level Rise," *Nature* 517 (January 22, 2015): 481–84, doi:10.1038/nature14093.

19. Robert William Sandford and Kerry Freek have produced a first-rate booklet on these floods called *Flood Forecast*, published by Rocky Mountain Books.

Chapter 11

1. September 29, 2014: On the path past 9 billion...

2. de Villiers, *Our Way Out.*

3. Mark Tran, "Global Overpopulation Would 'Withstand War, Disasters and Disease,'" *Guardian Weekly*, November 7, 2014.

Chapter 12

1. For a good catalogue of water conflicts, see Peter H. Gleick, *The World's Water*, vol. 8, 162.

BACK TO THE WELL

2. Mark Zeitoun and Jerome Warner, "Hydro-Hegemony — A Framework for Analysis of Trans-Boundary Water Conflicts," *Water Policy* 8 (2006): 435–60. My emphasis. Reprinted by permission.

3. Sandra Postel and Aaron Wolf, "Dehydrating Conflict," *Foreign Policy*, September 18, 2001.

4. ICA (Intelligence Community Assessment) ICA 2012-08, February 2, 2012, p. 1.

5. OECD, "Water and Violent Conflict," May 26, 2005.

6. I have borrowed the term "flashpoints" from the Circle of Blue website, which has a thorough and reliable database of potential conflicts and potential resolutions. Some of my material was sourced from this database.

7. Alon Tal and Yousef Abu-Mayla, "Gaza Need Not Be a Sewer," *New York Times*, December 2, 2013.

8. Asserted in letters from Weizmann, then head of the World Zionist Organization, to various British government officials in 1919 and 1920, and in a letter to David Lloyd George. There are multiple citations, including Hussein Amery, "The Litani River of Lebanon," *Geographical Review* 83, no. 3 (July 1993).

9. "Lack of Sufficient Services in Gaza Could Get Worse without Urgent Action, UN Warns," UN News Centre, August 27, 2012, http://www.un.org/apps/news/story.asp?NewsID=42751&Cr=Gaza&Cr1=#.VCwaMitdVkE.

10. Rana F. Sweis, "A Parched Jordan Places Hopes in Reservoir," *International Herald Tribune*, November 28, 2012.

11. Headline in *Economist*, March 9, 2013.

12. See Felicity Barringer, "Groundwater Depletion Is Detected from Space," *New York Times*, May 30, 2011.

13. See http://www.nasa.gov/grace and http://www.csr.utexas.edu/grace.

14. Quoted in "Less fertile crescent," *Economist*, March 9, 2013.

15. Francesco Femia and Caitlin Werrell "The Arab Spring and Climate Change," Climate and Security Correlations Series, Center for American Progress, February 28, 2013.

16. These quotes and the three in the following paragraph from "Foreign Ministers Try to Quell Tensions over Ethiopia's Plans to Divert Blue Nile in Controversial Dam Project," Al Jazeera, June 18, 2013.

17. See Adam Hafez, "How Yemen Chewed Itself Dry," *Foreign Affairs*, July 23, 2013.

18. Nasir Jamal, "Sound Bytes: Scrapping the Water Treaty Is No Solution," *Dawn*, October 3, 2014.

19. A UPI dispatch, quoted by Palash Ghosh in "What Are India and Pakistan Really Fighting About?" *International Business Times*, December 27, 2013.

20. Editorial, "Dark Times Ahead: Water Scarcity, *Dawn*, November 27, 2013. Also quoted by Ghosh, ibid.

21. See timesofindia.indiatimes.com/india/hafiz-saeed-blames-india-for-pakistan-floods-calls-it-water-terrorism/articleshow/42116443.cms.

22. For more on Mamata Bannerjee and the Teesta row, see Sougata Mukhpadhyay, http://ibnlive.in.com/news/mamata-row-clouds-pms-bangladesh-visit/181637-37-64.html.

23. This from the former Bangladeshi ambassador to the UN, Harun ur Rashid, in the Bangladeshi newspaper *Daily Star:* "Teesta Water Sharing: Some Hard Facts," October 11, 2014.

24. Teresa Rehman, "India's Brahmaputra River Cruises Come of Age," Al Jazeera, October 9, 2014.

25. Raushan Nurshayeva, "Uzbek Leader Sounds Warning over Central Asia Water Disputes," Reuters, September 7, 2012.

26. BBC, July 16, 2009, "Turkmenistan to Create Desert Sea," http://news.bbc.co.uk/2/hi/asia-pacific/8154467.stm.

27. Elizabeth Rosenthal, "In Spain, Water Is a New Battleground," *New York Times*, June 3, 2008.

28. "On the Table: Water, Energy and North American Integration," Munk Centre for International Studies, October 2007.

29. Anderson and Snyder, *Water Markets*, 196.

Chapter 13

1. *Nature* 514, no. 7 (October 2, 2014), doi: 10.1038/514007a.

2. Fishman, *The Big Thirst*, 301–2.

3. Sandford and Freek, *Flood Forecast*, 65–66.

Chapter 14

1. Peter Gleick, William Burns, Elizabeth Chalecki, Michael Cohen, Katherine Kao Cushing, Amar Mann, Rachel Reyes, Gary Wolff, and Arlene Wong, *The World's Water, 2002-2003* (Oakland, CA: Pacific Institute, 2002).

2. Gary Wolff and Peter H. Gleick, "The Soft Path for Water," in *The World's Water, 2002-2003*, chap. 1.

3. The top ten seawater desalination countries by online capacity are:

Saudi Arabia	9,170,391
UAE	8,381,299
Spain	3,781,314
Kuwait	2,586,761
Algeria	2,364,055
Australia	1,823,154
Qatar	1,780,708
Israel	1,532,723
China	1,494,198
Libya	1,048,424

The markets which are expected to see the fastest growth in desalination over the next five years are South Africa, Jordan, Mexico, Libya, Chile, India, and China, all of which are expected to more than double their desalination capacity. *Source:* International Desalination Association.

4. The Felber quotes are from an excellent piece by Julie Pyper: "Israel Is Creating a Water Surplus Using Desalination," *ClimateWire*, February 17, 2014.

5. Associated Press dispatch quoting *Haaretz*, "Israel's Desalination Program Averts Future Water Crises," May 31, 2014.

6. The Wald quotes too are from Pyper, "Israel Is Creating a Water Surplus Using Desalination."

7. Debra Kamin, "India Seeks Water Management Lessons from Israel," *New York Times*, June 12, 2013.

8. For his full speech, see the Prime Minister's website, http://www.pmo.gov.sg/media-release/speech-prime-minister-lee-hsien-loong-official-opening-tuaspring-desalination-plant.

9. Some of this from Edward Wong, "Desalination Plant Said to Be Planned for Thirsty Beijing," *New York Times*, April 15, 2014.

10. Ellen Knickmeyer, "Spurred by Drought, California Town Builds Desalination Plant," Associated Press, January 3, 2015.

11. NPR interview by Steven Cuevas of station KQED, "Southern California Better Prepared for Drought," February 10, 2014.

12. World Health Organization, "Safe Drinking-Water from Desalination," WHO/HSE/WSH/11.03.

13. World Health Organization, *Nutrients in Drinking Water* (Geneva: WHO, 2005), http://www.who.int/water_sanitation_health/dwq/nutrientsindw/en/index.html.

14. "Clean the Water: When It Is Muddied by Misinformation," January, 29, 2013, www.cwqa.com/_faq/misinformation.in c.pdf.

15. See Larry Greenemeir, "A Fine Brine: New Desalination Technique Yields More Drinkable Water," *Scientific American*, May 22, 2012.

16. "Desalination: A Useful Application May Have Been Found for Graphene: Improving Access to Fresh Water in the Developing World," *Economist* technology quarterly, June 2013.

17. Pyper, "Israel Is Creating a Water Surplus Using Desalination."

18. This from *Guardian Weekly*, October 14, 2014, in an article titled "Solar Energy: A Sunflower Solution to Electricity Shortage," by the *Observer*'s Robin McKie.

19. See the project's website, http://saharafore stproject.com/.

20. http://www.seawatergreenhouse.com/downloads/SFP%20Science%20mag%20Jan%20 2011.pdf.

Chapter 15

1. This and other Gleick quotes in this chapter are from the piece he co-wrote with Gary Wolff in Gleick, *The World's Water, 2002-2003*, called "The Soft Path for Water."

2. The report can be found at http://www.internationalrivers.org/files/attached-files/wwf32-crisis_0.pdf.

3. Quoted in Segerfeldt, *Water for Sale*, 117.

4. Quoted in Dan Fumano, "Outrage Boils over as B.C. Government Plans to Sell Groundwater for $2.25 per Million Litres," *Vancouver Province*, March 9, 2015.

5. "Ireland is a good news story, but not quite in the way the EU would like," *Economist*, November 8, 2014.

6. Speech by Alan Kelly, November 19, 2014. Full text at http://www.environ.ie/en/Environment/Water/WaterServices/News/MainBody,39549,en.htm.

7. For background, see http://www.bloomberg.com/bw/articles/2015-01-08/takadu-helps-israel-be-a-most-efficient-water-manager.

8. Quoted by Katia Moskvitch, then technology reporter for the BBC, Tel Aviv, "Intelligent Water Meters & Water Infrastructure Repair," December 1, 2011.

9. Ibid.

10. Wolff and Gleick, "The Soft Path for Water."

11. See www.compostingtoilet.org/faq/index.php.

12. C. Abegglen, M. Ospelt, and H. Siegrist, "Biological nutrient removal in a small-scale MBR treating household wastewater," *Water Resources*, January 2008: 338–46.

13. Some of these examples from Randall Archibald, "From sewage, added water for drinking," *New York Times*, Nov 27, 2007.

14. Ian Lovett, "Arid Southwest Cities' Plea: Lose the Lawn," *New York Times*, August 11, 2013.

15. Quoted in Samantha Larson, "Nestlé Doesn't Want You to Know How Much Water It's Bottling from the California Desert," Grist website.

16. https://www.nrdc.org/water/conservation/edrain/edrain.pdf.

17. http://water.columbia.edu/research-themes/data-analytics-and-multi-scale-predictions/braz il-allocation/.

18. Nilanjana S. Roy, "Leading a Push for Clean Water," *New York Times*, May 21, 2013.

Conclusion

1. Peter Haldeman, "In Los Angeles, a Nimby Battle Pits Millionaires vs. Billionaires," *New York Times*, December 5, 2014.

2. Jacques Leslie, "Deep Water: The Epic Struggle over Dams, Displaced People, and the Environment," *New York Times*, December 7, 2014.

Selected Bibliography

Anderson, Terry L., and Pamela Snyder. *Water Markets: Priming the Invisible Pump*. Washington, DC: Cato Institute, 1997.

Bakker, Karen. *Privatizing Water: Governance Failure and the World's Urban Water Crisis*. *Ithaca, NY*: Cornell University Press, 2010.

Barlow, Maude. *Blue Covenant: The Global Water Crisis and the Coming Battle for the Right to Water*. New York: New Press, 2009.

——. *Blue Future: Protecting Water for People and the Planet Forever*. Toronto: Anansi, 2013.

Barlow, Maude, and Tony Clarke. *Blue Gold: The Fight to Stop the Corporate Theft of the World's Water*. Toronto: McClelland and Stewart, 2003.

Black, Maggie, and Jannet King. *The Atlas of Water: Mapping the World's Most Critical Resource*. Berkeley: University of California Press, 2009.

Brown, Lester R. *Who Will Feed China? Wake-Up Call for a Small Planet*. New York: W.W. Norton, 1995.

Chartres, Colin, and Samyuktha Varma. *Out of Water: From Abundance to Scarcity and How to Solve the World's Water Problems*. London: FT Press, 2010.

Fishman, Charles. *The Big Thirst: The Secret Life and Turbulent Future of Water*. New York: Simon and Schuster, 2011.

Funk, McKenzie. *Windfall: The Booming Business of Global Warming*. New York: Penguin, 2014.

Gleick, Peter, ed. *Water in Crisis: A Guide to the World's Fresh Water Resources*. New York: Oxford University Press, 1993.

Gleick, Peter H., Lucy Allen, Juliet Christian-Smith, Michael J. Cohen, Heather Cooley, Matthew Heberger, Jason Morrison, Meena Palaniappan, and Paul Schulte. *The World's Water*. Vol 7. Washington, DC: Island Press, 2012.

Gleick, Peter H., Pacific Institute, Newsha Ajami, Juliet Christian-Smith, Heather Cooley, Kristina Donnelly, Julian Fulton, et al. *The World's Water*. Vol. 8. Washington, DC: Island Press, 2014.

Postel, Sandra. *Dividing the Waters: Food Security, Ecosystem Health, and the New Politics of Scarcity*. Washington, DC: World Watch Institute, 1996.

——. *Last Oasis: Facing Water Scarcity*. New York: W.W. Norton, 1997.

Robinson, Joanna L. *Contested Water: The Struggle against Water Privatization in the United States and Canada*. Cambridge, MA: MIT Press, 2013.

Rogers, Peter, Susan Leal, and Edward J. Markey. *Running Out of Water: The Looming Crisis and Solutions to Conserve Our Most Precious Resource*. London: Palgrave Macmillan,

2010.

Sandford, Robert William. *Cold Matters: The State and Fate of Canada's Fresh Water*. Victoria: Rocky Mountain Books, 2012.

Sandford, Robert William, and Kerry Freek. *Flood Forecast: Climate Risk and Resiliency in Canada*. Victoria: Rocky Mountain Books, 2014.

Sedlak, David. *Water 4.0: The Past, the Present and the Future of the World's Most Vital Resource*. New Haven, CT: Yale University Press, 2014.

Segerfeldt, Fredrik. *Water for Sale: How Business and the Market Can Resolve the World's Water Crisis*. Washington, DC: Cato Institute, 2005.

Index

rivers (*continued*)
    Padma 271
    Paraná 344
    Ravi 268
    Red 161
    Rhine 63–65, 69, 336
    Rio Conchos 282
    Rio Grande 22, 77, 174, 281–283, 290
    Ru 157
    Sacramento 185
    Salween 161, 251, 274, 344
    Seyhan 187
    Sieg 65
    Siuslaw 69
    St. Lawrence 53, 55, 60–61, 151, 344
    Sutlej 268
    Syr Darya 251, 276–277
    Teesta 271–274
    Tietê 51
    Tigris 250, 253, 260–262, 268
    Tonle Sap 48, 71
    Upper Drau 65
    Vaal 153
    Vakhsh 275
    Volga 155, 344
    White Salmon 159
    Willamette 69
    Yangtze 48, 50, 154–155, 161, 189, 344
    Yarmuk 251, 256–257
    Yellow 22, 48–50. *See also* Huang He
       River
    Yenisei 344
    Yukon 344
    Zambezi 78, 155
Robbins, Paul 325
Robinson, Joanna 92
Rocky Mountain Institute 311
Rodell, Matthew 35
Romania 66
Romm, Joe 222
Rosenthal, Elisabeth 230, 280
Rousseff, Dilma 19, 159
Russia 24, 46–47, 161, 169, 172, 180
RWE AG (company) 106–107, 113–115

S

S.A. Healy Company 109
Sabesp (company) 16, 17
Sabir, Munawar 270, 271
Sacramento CA 313
Safe Drinking Water Act 176

safe water 31, 104, 245, 295, 331
Saghir, Jamal 224
Sahara Forest Project 308, 310
Salina KS 213
Salini (company) 149
Salini Impregilo (company) 109
Salzburg Resolution on the Use of Inter-
    national Non–Maritime Waters 79
San Antonio TX 36, 87, 111
San Diego CA 186, 193–194, 199, 295, 301,
    325
San Francisco CA 54, 85, 95, 170, 188, 196
Sana'a Yemen 266–267
Sandford, Robert 291
Sandia National Laboratories 182
Sandoz (company) 65
Santarem Brazil 31
São Paulo Brazil 12, 15–16, 18–19, 25, 43, 51,
    116
Sarawak Malaysia 156
Saudi Arabia 23, 28, 172, 196, 244, 250, 258,
    262, 353
Sault Ste. Marie ON 200
Sawda Ghevra India 315
Scenic Hudson Preservation Coalition 61, 133
Schiermeier, Quirin 224
Schindler, David 59
Schlumberger (company) 173
Science-based Trials of Rowcrops Integrated
    with Prairie Strips (STRIPS) 213
Science (journal) 24, 67, 123, 309
Seager, Richard 223
sea-level rise 233–234
seas
    Aral 188, 216, 251, 276, 277
       destruction of 276
    Black 66, 261
    Bohai 48
    Caspian 251
    Dead 31, 47, 186–87, 190–192, 257, 260
    Galilee. *See* Lake Kinneret
    Mediterranean 31, 55, 188, 191, 256,
       297, 303
    Red 191, 192, 259
    South China 161
    Wadden 64
    Yellow 300
Seattle WA 12, 213
Segerfeldt, Fredrik 92
Semapa (company) 95, 96, 97
Senegal 125, 142, 216

BACK TO THE WELL